WP1391.
HD 31 ALE

Facilities Management

s to be returned on or before
the date stamped below

Facilities Management

THEORY AND PRACTICE

Edited by
Keith Alexander

London & New York

First published 1996 by E & FN Spon
Reprinted 1997, 1998

Reprinted 2000, 2001 by Spon Press
11 New Fetter Lane, London EC4P 4EE
29 West 35th Street, New York, NY 10001

Spon Press is an imprint of the Taylor & Francis Group

© 1996 E & FN Spon

Typeset by Acorn Bookwork, Salisbury
Printed and bound in Great Britain by T.J. International Ltd, Padstow, Cornwall

All rights reserved. No part of this book may be reprinted or reproduced
or utilized in any form or by any electronic, mechanical, or other means,
now known or hereafter invented, including photocopying and recording,
or in any information storage or retrieval system, without permission in
writing from the publishers.

British Library Cataloguing in Publication Data
A catalogue record for this book is available from the British Library

Library of Congress Cataloguing in Publication Data
A catalogue record for this book is available from the Library of Congress

ISBN 0–419–20580–2

♾ Printed on acid-free text paper, manufactured in accordance with ANSI/NISO
Z39.48-1992 and ANSI/NISO Z39.48-1984 (Permanence of Paper)

Library Hunters Hill
Marie Curie, Glasgow

Contents

List of contributors

Keith Alexander BSc (Hons), BArch, RIBA, ARIAS, MBIM, ACIArb, MIFMA, CFM
Director of the Centre for Facilities Management at the University of Strathclyde.

Craig Anderson PhD, ARICS, MIFMA
Course Tutor at the Centre for Facilities Management at the University of Strathclyde.

Ian Cooper BSc, BArch, PhD
Architect and Partner in Eclipse Research Consultants, Cambridge.

Roy Dyton BA, FCII, FIRM
Director of United Insurance Brokers Ltd, London.

Edward Finch BSc, MSc, PhD
Lecturer in the Department of Construction Management and Engineering at the University of Reading.

John Grigg MBA
Managing Director of Greenwood Grigg International, Wiltshire.

Oliver Jones BSc, MBA, ARICS, FAFM, MIFM
Managing Director of Symonds Facilities Management plc, London.

Ameen Joudah BArch, MSc(CABD), PhD, MIFMA
FM specialist and Project Manager with Hewlett-Packard, UK Real Estate and UK Support Services Department, Bracknell.

Alan Kennedy FBIFM
Director of Grampian Facilities Management Ltd, Aberdeen.

Paull Robathan MBA, FBCS, FInstD
Technology Strategist and a Director of Carlton Research, London.

Daniel B. Sweeney CFM
Principal of Daniel B. Sweeney Associates, Management and Consulting, London, and Visiting Professor at the Centre for Facilities Management at the University of Strathclyde.

Terry Trickett RIBA, FCSD, FBIFM
Director of Terry Trickett Associates Limited, London.

Bernard Williams FRICS
Partner in Bernard Williams Associates, Chartered Surveyors and Building Economists in Bromley, Kent.

Stewart Wood
Operations Director of the Government and Technology Business Unit of Procord Ltd, and is based at Harwell.

Foreword

In the last ten years we have seen the arrival of a new kid on the block. What gave rise to the appearance of facilities managers and where do they go from here? Their emergence can be traced to one thing – competition.

It is competition that drives the business world to embrace quality, to re-engineer processes and to look at the way in which work is done to bring about improvements in performance. The concept of customer satisfaction is now a hard-nosed reality for businesses wishing to survive, let alone to excel.

The turbulence created by these changes has provided the catalyst for the emergence of the facilities manager. Businesses focus on a reduction of overheads in an effort to improve their bottom line and their position relative to competitors. By similar means government seeks improved commercial success in the delivery of public services.

Downsizing, outsourcing and benchmarking are words in common usage in both sectors today. However, it would be all too easy to equate the discipline only with such words. I believe this would be a missed opportunity for facilities managers and their place in the business team. Currently they are recognized as the managers of non-core services, and charged with optimizing costs and improving the service levels in support of core business, as part of the drive to improve competitiveness.

Nevertheless the future environment in which they, and we, will work and live will be irresistibly shaped by advances in technology and communications. The former provide the means, the latter the opportunity, of participating in what will be nothing less than a revolution in the way people work and interact. Virtuality fast becomes reality.

Facilities managers have before them the opportunity to drive beyond improvements to existing ways of doing things to areas of cause rather than effect. It is often the case that they deal with the symptoms of business sickness, because of the way the business operates, rather than the root causes. To improve business performance is the goal – not to manage an overhead more effectively, but to seek to remove it entirely. Aggressive cost management should be given, as should quality and value, in the services provided.

It is the role of the facilities manager to challenge existing norms, to seek new methods of doing work in order to bring about marked and sustainable improvement in performance, not just within facilities areas but in the core of the business as a whole. Facilities managers should not be content only to manage services, rather they should seek involvement in finding solutions and innovative ways of addressing core business challenges.

I am often asked where such people will come from. They do exist, for there are already 'shakers and movers' within the profession. If more are to come forward, equipped for the changes ahead and with the agility and confidence to take facilities management to the boardroom as an integral part of business strategy, there is a need to invest in education and management development.

I trust you find this book an interesting introduction to the dynamics of the profession and I hope you will be sufficiently encouraged to invest in your future. Make no mistake, businesses of the future will demand the best of individuals and only the best is good enough.

You've got plenty of time: the future starts here.

Daniel B. Sweeney

Foreword

'Theory and practice' runs the title of this book, as if these were independent features rather than the two totally interdependent facets which should be present in every management operation. The Oxford Dictionary of course defines practice as 'action as opposed to theory' and theory as 'ideas or suppositions in general (contrasted with practice)', but it then qualifies this as 'a set of ideas formulated (by reasoning from known facts)...' implying that theory does have some kind of basis in reality!

In the context of synonyms the word pragmatic comes quickly to mind, implying a practical, down-to-earth corner-cutting approach to the root of a problem or activity, and usually inferring a disregard for 'the book' in the interests of achieving a workable solution quickly and cost-effectively. Of all modern management disciplincs facilities management (forced to grow too quickly out of an inadequately prepared base) is possibly the most plagued by this euphemism for bodging. People cut corners involving sound theory without properly understanding what it is they are bypassing and the risks inherent in such omissions.

As much a result of poor management as of under-resourcing, facilities managers constantly find themselves fighting fires, always reacting to adverse situations and rarely having the time to do things 'by the book'. Or is it the 'time to do things' that they lack or the 'time to find out the proper way to do things' for which they are inadequately educated? The danger inherent in the presence of a little knowledge has long been adopted as conventional wisdom yet still we have the second largest item of business expenditure – facilities – managed, by and large, by people who have received little or no formal training in facilities management. It is not surprising, therefore, that fire-extinguishers are ahead of textbooks on the practising facilities manager's shopping list.

Training seminars and degree courses for facilities managers are of course now increasingly available and the take-up is growing at a reasonably encouraging rate. Good quality reading matter is, however, in very short supply, especially that which introduces practitioners to good theory and novices to good practice. This work, despite the troubling innuendo of the title, does in fact fill exactly this gap, adding considerably to the thin body of available knowledge on the many and disparate facets of facilities management which it covers.

But who should read it and who should go on what courses and why? Not everyone responsible for service delivery management needs to understand space planning or how to use discounted cash flow, and the facilities task coordinators do not require in-depth knowledge of boiler maintenance or the cleaning of marble surfaces. So the dichotomy facing educationalists and publishers alike is finding out what their customers want (and need) to know to help them do their job better. This problem is compounded by an almost universal failure to distinguish between the different roles of sponsorship and delivery of facilities services and their quite different requirements with regard to information and management skills. Sponsors and providers alike will readily know what is for them and get early access to all the theory they need to make a valid practical contribution in their respective occupations, if educationalists define the role which their students fulfil (or aspire to) and structure the package of goodies accordingly.

There is much in this book of relevance to the sponsorship/intelligence facets of facilities management as well as that pertaining to the service delivery. But, before diving into the pages, the reader is strongly advised to work out which of the hats he or she is wearing, and if wearing more than one to look in the mirror to see if that is a sensible way of going about things. If they are not sure, they should do a retake holding a fire-extinguisher.

Bernard Williams

Preface

Facilities management is firmly established as an academic discipline in the higher education sector in the United Kingdom. Many universities offer facilities management courses and modules, most prominently in the built environment subject area. As the nature of facilities management has become more recognized, and as the role of the intelligent client has developed, greater emphasis is being given to strategic issues and to organization and management skills. Facilities management has developed from its technical base to become more of a management discipline.

This book is intended as an introductory reader for those interested in facilities management theories and their application to practice in the United Kingdom, as well as for those following facilities management courses. It serves as a prologue to its concepts, processes, tools and skills, and provides reference to sources of further reading.

The book is organized to reflect the structure of the Masters programme in Facilities Management offered at Strathclyde University. It introduces a strategic framework for the subject and raises issues for the organization and management of the facilities team.

These introduce a balance of generic management skills, including a core of quality, value and risk processes, focused on facilities operations and projects. Those operational skills for the delivery of particular facilities services cover the management of space, the environment, communications and the full range of services that support business effectiveness. Each chapter covers one module of the framework and contains an overview of the subject matter. This is followed by a keynote paper raising some of the main aspects of the subject as contributed by a visiting lecturer. A brief summary, and suggestions for further reading, concludes each chapter, and extended reading lists complete the book.

The Centre for Facilities Management would like to acknowledge the role played by its visiting lecturers, including those whose contribution is not marked by the inclusion of a chapter in the book, in developing a course which is industrially relevant and reflects the cutting edge of practice in the field. This book, and the full range of activities of CFM, would not be possible without the constant enthusiasm and total commitment of the core team to whom special thanks are due.

Keith Alexander

Editorial

In the world of business, decisions are driven by pressures to improve quality, reduce costs and minimize risks. This means that organizations must frequently undergo deep cultural changes involving all business processes, often with fundamental shifts in their organizational structures and working patterns. In this climate of change, quality of life and environmental issues are involved which together with issues of heritage and of conservation become part of the business agenda. Facilities management is at the heart of this process.

Facilities management is a key business discipline. Large corporations recognize its importance for business success. Smaller enterprises improve their performance by making better use of resources, often by forming networks and partnerships. Consultants and contractors in turn develop a variety of services to support this organizational effectiveness.

When times are difficult in business, when there is increasing corporate restructuring and growing uncertainty, facilities management becomes an area of substantial growth. Evidence of this can be seen as those professional bodies of the construction, hospitality and information technology industries increasingly recognize the opportunities that facilities management provides to their members to develop new value-added services, better integrated with an organization's business needs. Many relevant professional associations consolidate and work together to promote the professional status and role of facilities managers.

Facilities management consultants offer management and advisory services and undertake work on behalf of corporate and public sector clients on an agency or management contracting basis. Major contracting organizations offer contracted business services, either independently or in a 'total facilities management' package. Given the quest for improved effectiveness, especially when promoted by government policy, the market in both the public and private sectors expands rapidly.

In a business context where the raised expectations of clients lead to enhanced customer awareness, where economic, technological and environmental issues are affected by proliferating legislation, and where traditional discipline boundaries are challenged by information technology, facilities management can have its greatest impact. Many of these contexts also involve a redefining of the roles of purchaser and provider, and a move towards a contract basis for service delivery.

The informed buyer or intelligent client requires more management time at the interface with the user, more effective use of contracting out and less time administering inappropriate and cumbersome contracts. Effective quality systems, in which there is an appropriate balance between social, environmental, technical and economic issues, can release time to enable 'competition for quality', but who understands enough about the client's business that they can deliver quality facilities and so realize organizational effectiveness? Can such a person exist?

Market research has identified the need in facilities management for a kind of hybrid manager, a cross between professional manager and technical professional, combining the ability to make things happen with a level of technical understanding to enable facilities in organizations to be tuned to strategic needs. This requires the development of an appropriate balance of management and technical skills. Such a balance can lead to healthy and ecologically sound buildings in a business context, and resolve the relationships between organizations, the individual, the environment and the business community.

Strategic facilities managers are judged by their managerial capabilities rather than their technical competence. In many cases this means adding business, social and personal skills to a technical skills base. New groups of management skills, covering people, marketing, purchasing and contract, are involved. In an applied field like this it is vital that these skills are exercised in a practical, industrially relevant context.

The education and training of facilities managers, underpinned by research, must either be seen as part of general management development, or end up being marginalized. Facilities managers must have an understanding of facilities and services, recognize when further expertise is needed, contribute to strategic business planning and be in line for senior management posts in an organization.

Occupational competency-based management standards for facilities managers are essential. An open, individual, corporate, continuing system for such standards has to register a demonstrable ability to manage at junior (operational), middle (tactical) or senior management (strategic) levels. The aim is a mix of technical competence, language skills, cultural sensitivity and, at the top, a controlling vision that can bind activities together in many different countries.

The need for such able and experienced managers is strong and widespread. To meet the challenges of the next century the profession must organize in such a way as to be responsive to these needs. Professional associations in facilities management will flourish if they develop an open and positive approach to serving the needs of the consumer. United they could foster a strategy to:

- define standards for quality of performance and service;
- provide full support to practitioners by establishing a collective knowledge base and providing up-to-date information;
- provide guidance to the educational sector to support the development of modular courses at undergraduate, postgraduate and post-experience level;
- provide the means for an open exchange of experience and skills;
- create the conditions to promote advancement of the discipline, receptive to new ideas and supportive of a breadth of vision.*

Advancement of the profession hinges upon the assurance of quality and upon the value added by the service. The agenda for a professional quality system includes creating a ladder of educational opportunity and establishing qualifications, promoting the continuous development of skills and experience, and creating and maintaining an accessible collective database.

By establishing this quality system the profession can achieve a status and recognition that will provide an influence in business strategy and a strong voice in environmental politics. Such developments require close collaboration between academics, practitioners and researchers on a pan-national basis. The mechanisms should be created to support the processes of professional development through research, education and training.

If facilities management is to be acknowledged as a profession with its own rigorous discipline, it needs to sow the seeds for a strategy and infra-structure to promote development. 'Centres of excellence' should be created and linked into a network, to provide the focus for all this.

The University of Strathclyde, through the Centre for Facilities Management, has been involved in facilities management education in the United Kingdom since the middle of 1980. The first short courses were jointly presented with DEGW in 1984. The MSc and Diploma in Facilities Management has been running since 1987, with the aim of graduating high quality, innovative thinkers with a balance of managerial and technical skills and understanding, capable of contributing to the development of facilities management research and practice. Since 1994, CFM has consolidated its role and diversified its courses still further. Its aims and objectives are met by providing a mix of quality learning opportunities, supported by quality learning materials, information and facilities with the support and guidance of well qualified core staff. It is working to create qualifications that provide a ladder of opportunity to bridge from functional to general management.

To ensure it is fully aware of and can anticipate change, and to maintain standards, CFM has an Advisory Board, comprising representatives from all sectors of the industry. It makes visiting appointments to benefit from the experience and expertise of the leading practitioners and researchers in the field and it has strong working relationships with prominent companies. Professional associations in the FM field are also represented.

The Centre is a focus for research and discussion and seeks to create an atmosphere of enquiry and debate. It also provides the opportunity to meet leading experts from education, research and practice. CFM staff are well qualified, experienced professionals dedicated to supporting learning needs. They are all active in research, provide services to some of the leading companies in the UK, and bring up-to-date knowledge and recent experience to its courses. It is in the forefront of facilities management education in the United Kingdom and has a flourishing international reputation for its work. Its aim is to be recognized as a 'centre for excellence' in the field.

*Professional qualifications and undergraduate and postgraduate courses are increasingly available. Associations have been formed to promote facilities management and to represent the interests of members in the United States (IFMA), Japan (JFMA), Australia (FMA) and in Europe in the United Kingdom (BIFM), the Netherlands (NEFMA), Germany (GEFMA), Denmark (DFM), Hungary (HUFMA) and Finland (FIFMA). The associations, academic institutions and research organizations in Europe collaborate through the European Facilities Management Network (EuroFM).

<table>
| Facilities management | 1 |
</table>

Facilities management

1

OVERVIEW

Facilities management – 'the process by which an organization delivers and sustains support services in a quality environment to meet strategic needs.'   Centre for Facilities Management

When organizations undergo the kind of deep cultural shifts that involve all their business processes, they inevitably make fundamental changes to their organizational structures and working patterns. In such circumstances, issues concerning the quality of life and the environment need to be addressed and heritage and conservation become an important part of the business agenda. Nevertheless, business pressures to improve quality, reduce costs and minimize risks will continue to drive decisions.

Organizations of all kinds in different economies around the world recognize that the rising costs of occupying buildings, providing services to support business operations and improving working conditions are important factors in profitability. Success can depend upon reducing the costs of being in business. As buildings become more complex and house more technology, user expectations rise and the pressure on them to perform increases. Increasing legislation to ensure health, safety and welfare as well as to protect the environment has added new responsibilities on companies to manage the workplace.

Facilities management is the process by which an organization ensures that its buildings, systems and services support core operations and processes as well as contribute to achieving its strategic objectives in changing conditions. It focuses resources on meeting user needs to support the key role of people in organizations, and strives to continuously improve quality, reduce risks and ensure value for money. It is clearly an important management function and business service. Major organizations world-wide are using it as part of their strategy for restructuring to provide a competitive edge. It can also ensure that buildings and support services improve customer responsiveness and contribute to business objectives.

The scope of the discipline covers all aspects of property, space, environmental control, health and safety, and support services, and requires that appropriate control points are established in the organization. The facilities plan will set out these policies and identify corporate guidelines and standards. It will describe the organization, its structure, procedures and responsibilities. Facilities management policies lay out an organization's response to vital issues such as space allocation and charging, environmental control and protection, direct and contract employment. The policies will set a direction for the organization and establish the values of and attitudes towards the facilities users – the corporation, its operating units and customers, individual employees and the public.

Facilities management is relevant to organizations in all sectors and in developed and developing countries. However, differences in culture and management style must be recognized when delivering the operating environment and services needed for business effectiveness in a particular context.

Keynote paper

Facilities management: a strategic framework

Keith Alexander

INTRODUCTION

This paper aims to set the scene. Facilities management is first and foremost about organizational effectiveness. The decisions taken about facilities are business decisions. The business case for developing facilities depends on an understanding of the potential of facilities for creating quality working conditions to support key activities. Effectively planned facilities and quality support services can create significant business returns. As competition intensifies, and as change accelerates, many leading organizations are re-evaluating the contribution that facilities make to business success, recognizing the business consequences of poorly managed facilities and searching for value that can be added through effective planning and management.

THE BUSINESS AGENDA

Strategic decisions in these organizations, including those about adapting to the changing market, restructuring and investment opportunity choices, will be made by their chief executives and the managing directors of their operating units. It is at board level that conditions are set within which facilities are operated and developed. Physical resources and support services may be treated as non-core activities and so not figure in boardroom considerations. However, sites and buildings project the identity of the organization and those who deliver support services are often the first line of contact with customers. If the lasting impressions of an organization are formed in the first five minutes contact, these 'moments of truth' are vital. Facilities include the space, environment, communications and services that enable the achievement of key business objectives. They may be looked at in one of two different ways in any organization – as property, the fixed assets that appear on the balance sheet, or as an administrative overhead, a business expense to be minimized. They are often overlooked as a source of competitive advantage.

Even where facilities management has been introduced it may be seen as little more than collecting responsibilities under a convenient functional head. Buildings may be treated as liabilities rather than assets – a drain on scarce resources. Similarly general services may be seen as a support function – an overhead cost. Neither view enables organizations to use the full potential of buildings, communications and support services in improving business performance. Facilities need to be offered as a service, and provided, for a charge, to the operating units. Their scope may include business services, e.g. legal, accounting and administrative as well as space, environment, information and other support activities. However, many companies still view the provision, operation and maintenance of facilities from a technical and project base rather than from a business perspective.

There is ample evidence of poor property management – research studies by the Oxford Brookes and Reading Universities concluded that 'property, though a key asset for operational organizations, is ill-considered and poorly managed by most'. They identified examples of good practice and raised the questions that organizations should address if they are to begin to make their property work for them. Researchers have written extensively on the management of operations and performance in relation to property over the past decade. There is also evidence of inadequate service delivery and recognition of the adverse consequences of undervalued and under-utilized facilities for corporate performance. Few organizations in fact fully recognize the contribution that facilities can make to the business, and few can identify the opportunities they provide.

There are many examples of the positive impact that effective facilities management can have on success as measured by key business indicators such as cost/income ratios, product lead time and process improvement. Corporate attitudes toward facilities must be towards their management, either as a business service or as a company asset. They must be developed as corporate assets to add value to core business activities and managed to offer service quality in support of business operations. Separate initiatives to improve facilities, to sharpen the corporate image, to enhance operational efficiency and increase cost-effectiveness may make a marginal difference to company performance. Only by being tuned to business objectives and managed to a strategic plan can they be organized to encourage innovation and enterprise. This is a strategic role for organizations – to develop policy, contribute to strategic planning, negotiate service levels and arrange for the delivery of quality facilities.

A CONTEXT OF CHANGE

Many organizations restructure to create more appropriate relationships in what are in effect 'new organizations'. Others review critical success factors and seek to improve business processes. Wholesale corporate re-engineering concentrates on the improvement of all business processes.

Surveys conducted by Frank Duffy and colleagues as part of the Responsible Workplace Study (Butterworths, 1993) identified the pressures for change (technology/telecommunications and organizational structure), workplace issues (location, servicing, layout, security/access) and decisions that organizations considered most important for the future (responsive adaptation for user needs). This particular study focused on innovative organizations introducing advanced information technology and devising organizational structures to cope with change. It also dealt with the search for ways of increasing productivity and of using time more effectively as well as responding to expectations for a safer working environment and taking environmental responsibilities seriously.

There are complaints that 'the new gurus (only) talk about vision or strategic intent of companies'. Charles Handy, in the Michael Shanks Memorial Lecture 'What is a company for?' (*The Royal Society of Arts*

Journal, March 1991, pp. 231–341), could be seen as an example of this. However, in many cases, architects and suppliers of physical space are not given much opportunity to link the process of designing office space with such strategies. Business strategists need to simultaneously shape the form of their organizations as well as the working environment through which that future can be achieved. Paull Robathan, presenting his paper *The Intelligent Perspective* at the EuroFM Conference in Rotterdam in September 1992, has suggested that facilities managers are best placed in organizations for this kind of business re-engineering.

However, the role of facilities management is broader than the design of the product and production of the physical workspace. It entails the integration of people, technology and support services so as to achieve an organization's mission. It is concerned primarily with the quality of service to all stakeholders in the organization. These management issues and their implications for the organization – concepts of excellence, teamwork, total quality and service – must be clearly recognized and understood if facilities are to be effectively managed to support the corporate mission and objectives. They have a fundamental impact on the workplace and upon delivering the facilities. In turn, the development of facilities in an organization can act as a catalyst to cultural change.

ORGANIZATIONAL EFFECTIVENESS

The effectiveness of organizations may be measured in different ways, depending on their mission, prime goals and objectives, as well as the relative influence of the stakeholders – those groups with an interest in it:

- customers – customer care/customer first – 'moments of truth';
- employees - investing in people;
- shareholders and creditors;
- collaborators and suppliers;
- the community, and society;
- government.

Stakeholder analysis is used by some firms to identify and classify their expectations.

Organizational effectiveness may be described as: maintaining commitment amongst the members of the organization, communication amongst operating units, projecting a positive and responsible image, enabling change and improving productivity. Facilities have the potential to contribute and it is important to identify (and measure) the extent that they support, or can be adapted to, the changing needs of organizations, and contribute to productivity, profitability, service and quality.

Companies are revisiting basic questions about the purpose of the organization, asking 'what business are we in?' 'who are our customers?' and developing a core business philosophy. Charles Handy suggests that the 'principal purpose of a company is not to make a profit – full stop. It is to make a profit in order to continue to do things or make things, and to do it even better and more abundantly. To say that profit is a means to other ends and is not an end in itself is not a semantic quibble, it is a serious moral point.' Handy's point is that, increasingly, companies will need to think about their social environmental performance, and how their products or services benefit a wider social good.

ORGANIZATIONAL DYNAMICS

As organizations respond to the business climate and adapt to legislative, political, social and economic influences, their facilities needs change. All organizations are dynamic and undergo differing levels of change, much of it structural, to adapt to an evolving business environment. With the accelerating pace of such change, balancing demand and supply in managing the operational environment has become a vital element in strategic planning. Privatization with restructuring is a good example. More and regular change, to adapt working patterns to give continuity of workflow by forming and reforming project teams, is another. As the rate of change in an organization increases, so does the importance of managing facilities as a factor of production.

Organizational change affects the size and shape of organizations, working patterns and customer and supplier relations. These factors need to be constantly under review. Understanding the changing organization is similar to Mark Twain's description of the process of mapping the Mississippi – a constantly changing landscape. Continuity of production or service with reliable support and no disruption is important in all organizations. This entails assessing the business risks associated with the operating environment and having a plan to manage risks and deal with contingencies. The consequences of interruption could be disastrous. All business resources must be effectively utilized to ensure competitiveness. The significance to financial viability of the rising costs of occupying buildings and delivering support services is increasingly recognized. Facilities are an organization's second largest expense and can account for as much as 15% of turnover. The pressure to reduce costs is usually the main driver of a company's decisions about facilities. They are also the largest item on the balance sheet, typically over 25% of all fixed assets, so it is important that they are managed to add value to the organization.

Many organizations realize that human resources are the key to success. Demographic trends suggest that, in Europe, a highly skilled workforce capable of using the potential of advanced technologies may well be in short supply. More attention has to be paid to ensure that companies can attract and retain the right quality of staff and a package of working conditions, including the facilities to support the quality of working life, will be a prerequisite for organizations who wish to compete. Increasing demands are being placed on buildings to provide an improved working environment with new information technologies. New approaches are needed as health and safety and environmental legislation requires managements to comply with improving standards.

It is clear from all this that facilities management is not simply the operation and maintenance of buildings, the provision of cleaning services or the recording and rearranging of furniture in offices. From the highest level, in any organization, the strategic aspect of planning for the effective provision of services offers opportunities for economy, efficiency, effectiveness and competitive advantage.

FACILITIES MANAGEMENT PROCESSES

In these conditions, organizations must ensure that they are responsive to change, and able to take advantage of technological innovation. Facilities management is the application of the total quality techniques to improve quality, add value and reduce the risks involved in occupying buildings and delivering reliable support services. Such an approach is required to provide and sustain an operational environment to meet the strategic needs of an organization. An ambience of quality can ensure that core business processes are well integrated and supported in an operational environment – the workplace. The process is cyclical and relates needs to a result that can be tested against user satisfaction with the service:

- space – adapted to changing needs and effectively utilized;
- environment – to create healthy and sustainable working environments;
- information technology – to support effective communications;
- support services – to provide quality services to satisfy users;
- infrastructure – to provide appropriate capability and reliability.

Organizations should have a clear strategy and well developed policies for facilities management embodied in a facilities plan, and should establish a single point of responsibility. The way in which facilities are organized in relation to central functions and to other operating units will determine the extent to which facilities support strategic needs. Value is added to an organization at the workplace through the provision of services in the most efficient and effective way: by the development and management of quality managed systems, through the establishment of guidelines and service levels and, at the policy level, through the development of a strategy and a framework within which to deliver services.

The facilities planning process identifies user needs and agrees service levels as the basis for designing the service. Effective planning of facilities will ensure that they are 'work-shaped' and support work processes. The services that are required can then be defined, specified and delivered. Quality systems ensure that all services are delivered to the required quality, provide value for money and minimize risks to the organization.

Such customer response is a key indicator of facilities performance. Most leading organizations are engaged in a process of identifying their performance in key business areas as part of their quality initiatives. Increasing attention is now being given to the performance of facilities as a factor of production and as a business resource. Financial indicators still predominate but increasing interest is being shown in social and environmental indicators. As well as tracing historical performance, companies will seek to compare their performance against other similar organizations and against the 'best of breed – the benchmarking process. Such comparisons depend upon the availability of standard ways of measuring and representing performance and cost information.

FACILITIES MANAGEMENT ORGANIZATION

The level of responsibility and degree of autonomy provided to the facilities enterprise and the commitment and encouragement they receive to improve quality will determine the value that can be added through facilities management. Appropriate relationships should be established within which people are empowered and decisions about the use of facilities are taken closer to the customer. A single point of responsibility should be established as the informed buyer of services, concentrating management resources on the strategic role. A facilities director or senior manager will be primarily responsible for tuning facilities to the needs of the business.

The facilities team must be organized to achieve strategic control. The model of an enabling organization will promote more effective business-like management, which pays more attention to customer requirements and value for money. Responsibilities for policy formulation and service delivery are separated. This enables senior managers to concentrate on the core responsibilities of setting priorities and standards and finding the best ways of meeting them. Negotiating contracts for services makes managers accountable for performance against specified standards and allows them to achieve the best value for their resources.

Many organizations now distinguish between the roles of purchaser and provider, and have moved towards the contract as a basis for service delivery. Organizations can release management time through using consultants and contractors. Third-party facilities management, outsourcing options and contracting-out can all play their part. They also have to resolve issues concerning managing the facilities enterprise and privatization. The evolution of a facilities management organization can be traced from introduction and early awareness to consolidation and maturity as an autonomous business unit.

Facilities management policies lay out an organization's response to vital issues such as space allocation and charging, environmental control and protection, and direct and contract employment. The policies will set a direction for the organization and establish the values and attitudes towards the facilities users – the corporation, its operating units and customers, individual employees and the public. The facilities plan will set out these policies and identify corporate guidelines and standards.

THE INTELLIGENT CLIENT AND INFORMED BUYER

Organizations require the effective management of customers, assets and level of service – three related aspects of any facilities operation. Managing customer expectations and meeting their requirements implies a total quality approach to developing and operating buildings and delivering support services to contribute to achieving business objectives. The full costs of quality must be recognized and these costs can be reduced with appropriate attitudes, improved processes and effective teamwork, all geared towards the customer.

With service level agreements and service contracts there have to be defined standards of performance and responsibilities for meeting them. If specific responsibilities, such as security, are devolved, this must be planned and optimized.

Effective managers will create more management time at the interface with the user. The development of effective quality systems will release valuable management time spent administering inappropriate and cumbersome contracts. These systems will provide choice to users of the service, will promote 'competition for quality' and enable effective use of contracting-out. An informed buyer assesses needs, agrees service levels and purchases services to meet them. The facilities organization must enable managers to focus on buying the best standard of service achievable within an agreed budget.

Facilities also represent considerable value as the fixed assets of an organization. The physical assets of an organization need to be managed effectively to ensure their value is realized and their repair, maintenance and replacement is planned.

PROFESSIONAL MANAGEMENT SKILLS

The business skills of facilities managers must be developed in the strategic context outlined above. A key to the success of facilities management, and to its development as a professional discipline, lies in achieving the appropriate balance of general management and technical skills for this strategic role. Leaders are required for creating the vision, for making the right choices for the future and managing complexity.

Strategic facilities managers will be judged by their managerial capabilities rather than their technical competence. In many cases this will mean adding business, social and personal skills to a base of technical skills. A 'hybrid manager' with a balance of technical and managerial skills will understand the nature of problems and make things happen.

The successful manager needs the imagination to foresee the consequences of change and the power to effect change. Seen in this way, facilities management is a core management activity and requires the active involvement of facilities managers in managing the mission, the business and the assets. They should have a vision of the service that will meet the business need, but also need the management skill to build and lead the facilities team to deliver to it. To be effective he/she must have the authority and be prepared to accept full accountability for all actions so that they not only manage well, but will be viewed as leaders within the organization. This may require radical personal and professional adjustment.

The ability to communicate is a key facilities management skill. Effective feedback mechanisms will ensure that the organization remains aware of the users' views at all times. The service provider team must be organized and motivated to deliver quality services for all customers through well managed processes. New purchasing and contracting skills, including a strong purchasing capability, will be required. A service manager will ensure that the specified level of services is procured and delivered to the appropriate quality, on cost and on time.

The nature of facilities management and its required skills are perhaps best described through reference to typical job advertisements. After declaring their unrivalled commitment to people and the workplace, organizations seek individuals to enhance their reputation through their attitudes and actions. Advertisements lay out the scope of the post and identify responsibilities to 'plan, create and maintain an internal and external working environment conducive to the safe, effective and efficient carrying out of all business activities within the high volume (business) facility'.

It is envisaged that this would be achieved through the proactive leadership of small highly trained teams and by the coordination of all contractor relationships. It is suggested that what would set applicants apart from others would be the ability to create an environment which matched a company's open, non-hierarchical organization and a belief that the right facilities would result in a competitive advantage. The task would be to set the standard for facilities control across a business and a positive 'can do' approach should be mirrored by the team and manifest itself by working effectively through others.

CONCLUSIONS

For the organization, facilities management means:

- creating a facilities policy that expresses corporate values;
- giving the authority to the facilities business unit to improve service quality;
- developing facilities to meet business objectives;
- recognizing the value that facilities add to the business.

For the facilities management organization, the strategic role entails:

- formulating and communicating a facilities policy;
- planning and designing for continuous improvement of service quality;
- identifying business needs and user requirements;
- negotiating service level agreements;
- establishing effective purchasing and contract strategies;
- creating service partnerships;
- systematic service appraisal – quality, value and risk.

For the business manager, facilities management implies a balance of skills:

- leadership;
- technical understanding and management know-how;
- personal and interpersonal skills;
- purchasing and contracting skills.

SUMMARY

A number of key factors have promoted the growth of facilities management – conditions of change, the increasing complexity of buildings and systems, pressures to reduce costs and improve competitiveness, rising user expectations regarding the quality of working life and increasing levels of legislation.

Challenges to the conventional ways of describing these factors suggest a new corporate reality. Can the corporate identity so easily be represented by the products, places and people in today's business world? The diversity of the new corporation, with 'invisible' products, dispersed workplaces and many ways of working, suggests not. This creates new challenges for those responsible for facilities and the corporate identity. Any physical environment which ignores the design implications of this diversity can quickly become obsolete.

The office is becoming less important in organizational terms as people do not travel to work as much and there are more temporary and mobile staff. This means there will be different ways of building communities. People will have a different allegiance towards their employers. In computing, for example, people who have been with the same company for six years or more are becoming rare. Instead, they will have other allegiances, with professional societies for instance.

The concept of workplace ecology includes a consideration of the physical, social, environmental and administrative setting for productive activity. Management and design must consider the broad dimensions of workplace ecology to create conditions in which all needs can be satisfied and objectives fulfilled.

BIBLIOGRAPHY

Essential reading

Alexander, K (ed.) (1993) *Facilities Management 1993*, Hastings Hilton, London.

Alexander, K (ed.) (1994) *Facilities Management 1994*, Hastings Hilton, London.

Alexander, K (ed.) (1995) *Facilities Management 1995*, Blenheim Business Publications, London.

Becker, F. (1990) *The Total Workplace: Facilities Management and the Elastic Organization*, Van Nostrand Reinhold, New York.

Recommended reading

Hammer, M. and Champy, J. (1993) *Re-engineering the Corporation: A Manifesto for Business Revolution*, Nicholas Brealey, London.

Handy, C. (1989) *The Age of Unreason*, Business Books, London.

Kanter, R.M. (1989) *When Giants Learn to Dance: Mastering the Challenges of Strategy, Management and Careers in the 1990s*, Simon & Schuster, New York.

Pascale, R. (1990) *Managing on the Edge*, Penguin, Harmondsworth.

Waterman, R.H. (1994) *The Frontiers of Excellence: Learning from Companies that Put People First*, Nicholas Brealey, London.

Zuboff, S. (1988) *In the Age of the Smart Machine: The Future of Work and Power*, Basic Books, New York.

Further reading

Becker, F. (1989) *Corporate Facilities Management: An Inside View for Designers and Managers*, McGraw-Hill, New York.

Brand, S. (1984) *How Buildings Learn*, Viking Penguin, London.

Campbell, A., Devine, M. and Young, D. (1990) *A Sense of Mission*, Economist Books, London.

Carlzon, J. (1987) *Moments of Truth*, Ballinger, New York.

Cassells, S. (1993) 'Power and imagination: the business of facilities', in Alexander, K. (ed.) *Facilities Management 1993*, Hastings Hilton Publishers, London, pp. 46–9.

Duffy, F. (1989) *The Changing City*, Bulstrode Press, London.

Duffy, F. (1992) *The Changing Workplace*, Phaidon, London.

Duffy, F., Laing, A. and Crisp, V. (1993) *The Responsible Workplace*, Butterworth Architecture, Oxford.

Goumain, P. (1989) *High Technology Workplaces: Integrating Technology, Management and Design for Productive Work Environments*, Van Nostrand Reinhold, New York.

Kotter, J. and Hesketh, J. (1992) *Corporate Culture and Performance*, Free Press/Macmillan, New York/London.

Tompkins, J.A. (1984) *Facilities Planning*, McGraw-Hill, New York.

<table>
<tr><td>**2**</td><td></td></tr>
</table>

2	# Organization and management

OVERVIEW

One of the most dramatic requirements associated with increasing responsiveness is to shift the organization's entire 'way of being' from a vertical (hierarchical) to a 'horizontal' (fast, cross-functional co-operation) orientation.   Tom Peters

A developing facilities management role is defined by its relationship with the core business and its success is measured by the support it provides in achieving key business objectives. Hence, facilities are not only strategically planned but have a clear relationship between their development and operation and the business need. In the climate in which business operates, the primary organizational need is for greater adaptability. Included in this is the need to anticipate differential rates of change, to accommodate expansion and contraction and to enable a rapid response to business opportunities. For this, management decisions need to be taken in real time.

Established organizations in the public and private sector in the United Kingdom, which have traditionally emphasized stability, status and position, are undergoing extensive programmes of change and corporate restructuring. Resulting pressures lead to a reduced headcount and closely controlled costs. Hitherto, morale has been linked to security, routine and stability in job tasks, but future demands are likely to focus on independence and creativity.

As organizations develop they take a different form with different working patterns. The indications are that, as management gurus suggest, these 'new organizations' are flatter without formal hierarchies, with decentralization and devolution of responsibilities, and with an end to long-term, secure careers for corporate managers. Organizations tend to evolve towards this new model as they seek to improve competitiveness. Successful companies and public bodies cultivate corporate cultures that emphasize innovation and change. They organize around work processes rather than around formal hierarchies and prescribed functions. They build around relationships of trust.

Many organizations distinguish between the roles of purchaser and provider, and adopt the contract as a basis for service delivery, which means that they can release management time through using consultants and contractors. The issues of third-party facilities management, outsourcing options and contracting-out are central, and standards of performance and responsibilities for meeting them are defined by service level agreements and contracts.

The value added by business services, particularly those involved with facilities management, is created by designing relationships that respond to these conditions and provide effective support for the achievement of business objectives. These relationships release management time and concentrate resources on the core business activities. Appropriate relationships are established within which people are empowered and decisions about the use of facilities are taken closer to the customers. The level of responsibility and degree of autonomy provided to the facilities enterprise, and the commitment and encouragement it receives to improve quality, determines the value that can be added.

Keynote paper

Transforming organizational life by design
Terry Trickett

INTRODUCTION

A determination to find new ways of improving the quality of working life for people at work can be identified as the single most important cause of change in organizations. As social psychologist W.G. Bennis observed: 'In every age there is a strain toward organizational form which will encompass and exploit the technology of the time and express its spirit' ('Changing organizations', *Journal of Applied Behavioral Science*, Vol. 2, No. 3). Inevitably, all organizations are concerned in this constant struggle to find new and more appropriate organizational forms but, too often, it occurs without an equivalent degree of effort being directed towards making changes in the working environment. The links between radical organizational change and the way organizations occupy space are not always made.

There can be no justification for ignoring or underestimating the support that can be provided by a satisfactory and responsive working environment. However, it has to be admitted that, from 1949 onwards, when Elton Mayo wrote *The Social Problems of an Industrial Civilisation*, most management theorists contributing to the knowledge of organizational behaviour have tended to relegate the role of the working environment to a low level of priority. One interpretation of the Hawthorne experiments, for instance, was that physical changes had little effect on work productivity – an apparently good reason for paying little attention to the working environment.*

Only in recent years, with the growth of the facilities design and management movement, can signs be seen of 'a fragile bridge being built between the social sciences and design skills' – a point well made by Franklin Becker in his book *Workplace* (Praeger, New York, 1981).

ORGANIZATIONS AND PEOPLE

The key to a successful organization is the extent to which each individual member feels that organizational objectives (and the way they are achieved) are significant to him or her. Likert, as evidenced in his book *New Patterns of Management* (McGraw-Hill, New York, 1961), is one of many management theorists who stress the need for encouraging a high degree of trust between superiors and subordinates. He rates the success of 'highest-producing managers' by their ability to create a general pattern of highly motivated cooperative members where control does not need to be exercised through authority:

> A highly motivated, co-operative orientation toward the organization and its objectives is achieved by harnessing effectively all the major motivated forces which can exercise significant influence in an

---

*Elton Mayo brought the analytical skills of social science to bear on a phenomenon that had been observed at the Hawthorne Works of the Western Electric Company in Chicago. Engineers, prior to Mayo's involvement, had found that in isolating two groups of workers and creating for one static lighting conditions and the other variable conditions, no significant differences in output could be observed. In fact, whatever was done with the lighting, production rose in both the groups. Mayo, in continuing these experiments, found that subsequent changes – shorter hours, rest pauses, incentive payment schemes – all produced an increase in output, not just in the 'test' room but also in the other monitor groups. Finally, in returning to the original conditions of work, output rose to the highest point yet

organizational setting and which, potentially, can be accompanied by co-operative and favourable attitudes.

Clearly, it is desirable that work is made more likeable and that active steps are taken to create job satisfaction at a personal level whenever possible. It is a tragedy that this is still such a difficult and seemingly unattainable aim in many commercial and industrial occupations. However, with regard to the conditions created for work, there is much that can be done. As a general rule, the buildings in which work takes place (and, in particular, office work) have changed little over the years; they give little sign that a revolution has occurred in the way work is regarded.

The interiors of many buildings show an ability to cope with well defined systems of organizational hierarchy and authority; they are conducive to predictability, order and precision. By comparison, very few demonstrate a capability of generating appropriate forms to match new and emerging management philosophies. Although a few signs of change have emerged, the 1990s are basically the beginning of a new phase of workplace design. A broad-based approach to environmental design is required, which can:

- allow a sense of order to emerge from diversity;
- encourage informal communication at all levels in an organization without usurping the sense of 'individuality' provided by personal space;
- act as an aid to creativity and, wherever possible, suit problem-solving, rapidly changing organizational systems;
- act as a salient reward where it is impossible to make all jobs intrinsically interesting;
- compensate for variations in individual skills and abilities;
- contribute towards people's self-esteem and their ability to perform tasks effectively.

These objectives recognize the social and psychological issues which must inevitably form a crucial aspect of environmental change. They defined the extent of the considerable contribution that can be made by the working environment towards achieving improvements in the quality of working life.

STRATEGIES FOR ENVIRONMENTAL CHANGE

By taking a broad-based approach to environmental change, facilities managers/designers have it in their power to transform organizational life. They can ensure that people at work gain ever increasing support from the environments in which they spend so much time. A series of eight strategies, based on experience gained generally but not exclusively in office design, can help achieve this.

Strategy One: Measure needs by evaluating people's wants

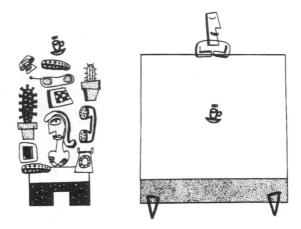

As Robert Sommer observed in *Personal Space: The Behavioral Basis of Design*: '...in the behavioural realm, the way buildings affect people, architects fall back on intuition, anecdote and casual observation' (Prentice Hall, Englewood Cliffs, NJ, 1969). He suggests, therefore, that designers need concepts that are relevant both to physical form and human behaviour – an approach which presupposes that it is possible to define and understand the determinants of people's behaviour at work in some depth. Behaviour which results from, formal organizational demands, informal activities and individual needs for self-fulfilment must all be taken into consideration to successfully uncover the 'unique pattern of activity' by which an organization generates its own success.

It is only by piecing together an organization from an examination of its constituent parts (at all levels within its hierarchy) that a balanced and accurate view of people's needs can be obtained. During this process of enquiry people's response to questions (e.g. what do you do? how do you do it? how do you relate to others during the process of your work?) may tend to reflect situations imposed on them rather than situations as they might be or should be. To overcome this 'communications gap', facilities managers and designers must be prepared to probe well below the surface of organizational life; this is where turbulent reality lies and people's feelings towards their workplaces become fully evident.

Strategy Two: Reflect management style by studying the ways people work together

Researches show that organizations have complex and variable styles of management influenced by the way work procedures are carried out (do people work individually or closely together in teams?), whether authority is concentrated at the top or dispersed, and their interpersonal style (do people relate to one another in a formal or informal way?). It is these factors which largely determine an organization's layout. (Layout can be defined as the explicit way in which groups of people are placed together in offices, the size and shape of the space they occupy, and the degree of separation that exists between them.)

A general change of direction can now be discerned in the way people work together. New attitudes towards management and the social climate in which work takes place mean that authority systems are tending to become more dispersed, working procedures more participative and systems of control more informal. As a result, few organizations require predominantly cellular space or, alternatively, open space with no partitioned divisions. Instead, organizations need mixed layouts where a subtle balance can be obtained between the technical needs of groups of people working together and a recognition of their social needs.

Strategy Three: Recognize social needs by removing barriers to communication

It is no good applying a company-wide planning system on the basis of how management wishes people to be. Instead, it has to be remembered that the operational patterns evident within organizations become increasingly complex as people working within a system of interlinked social/work groups endeavour to find ways of communicating effectively

with one another. As Likert says:

> ... individuals in organizations desire to achieve and maintain a sense of personal worth. The most important source of satisfaction for this desire is the response we get from the people we are close to, in whom we are interested, and whose approval and support we are eager to have. The face-to-face groups with whom we spend the bulk of our time are, consequently, the most important to us.
>
> (Likert, R. (1961) *New Patterns of Management*, McGraw-Hill, New York)

It follows that the way people are arranged in groups can either help or hinder their performance. It hinders when:

- screening placed between people causes them to work more against than with one another;
- too little space is provided at the individual workplace;
- provision for meetings at the workplace, or between group members, is either minimal or non-existent.

These factors act as barriers to communication. By contrast, group layouts help when they encourage interaction rather than preventing it (see Strategy Four).

Strategy Four: Encourage interaction by helping people to work in groups

People in groups need the help and support of fellow members to carry out their tasks effectively. The way these groups are formed – the physical configuration of work positions and the extent to which they are separated from one another – can help the performance of tasks and, at the same time, encourage interactions between people.

In planning groups, emphasis needs to be placed on the social networks which operate within them. These involve:

- meetings at the workplace;
- informal meetings within the group;
- opportunities to interact when performing combined tasks.

Shared activities of this type, combined with tasks carried out at individual workplaces, form the essential elements of group design. In bringing these elements together, designers should aim to think not only in spatial and planning terms but also in terms of the wants, values and expectations of group members. It will then become possible to select, from the wide range available, the most appropriate form of group layout to meet a specific set of technical and social needs. For the future, it can be anticipated that much more variety and stimulation will be introduced into offices by incorporating, within any single interior, a number of radically different types of group design.

Strategy Five: Improve efficiency and comfort by re-engineering workplace tools

The attraction of work itself is important because people's main concern is usually the quality of their immediate surroundings and the aids they are given to carry out their jobs. For this reason, the aim must be to provide easy access to equipment, materials and references, and 'support' tasks through the provision of work surfaces and storage units. It may be necessary to re-engineer these workplace tools. This approach banishes the idea that the office is a 'machine for working in'; instead it reintroduces the concept of providing tools (both varied and individual) which extend people's natural capacity for work. Once again, as in pre-

industrial days, people can become individually responsible, or responsible within a small group, for exercising judgement and assessment in order to produce quality.

Increasingly, in many jobs, people do have a choice of how they approach the tasks in hand. They can vary the amount of effort and energy they put into the job and they can vary the performance strategies they adopt. Research has shown, in L.W.L. Porter, E.H. Lawler and J.R. Hackman's book *Behavior in Organizations* (McGraw-Hill, New York, 1961) for example, that rewarding jobs of this type offer variety, autonomy, social interaction, use of skill and responsibility. All these attributes can be as much influenced by the surroundings in which work takes place as by the nature of the work itself.

Strategy Six: Influence people's attitudes by defining the surroundings in which they work

The extent to which individuals feel comfortable in their surroundings is influenced as much by psychological factors as the physical form of their workplaces. People in offices generally share an acute awareness of the personal space they occupy and the degree of security it offers. These are issues which demand very sensitive handling during the design process.

It is possible to imagine each individual as being enclosed in an invisible bubble which defines the extent of his or her individual territory. This bubble, in extending beyond the body, enables people to have a sense of control or ownership, over their immediate surroundings. It influences the way solitary activities are performed and the manner in which social interactions take place. The design of work settings, therefore, must take as much account of people's territorial needs as their needs for safety and comfort. It is by this means that facilities managers and designers have it in their power to strongly influence, even if they cannot determine, the attitudes and behaviour of people at work. People are 'shaped' by the interiors they occupy.

Strategy Seven: Increase variety by expressing people's differences

Above all else, it is necessary to find ways of recognizing the overriding importance of the individual – people are the organization. We must 'add value' to our interiors by giving close attention to the expression of individual differences.

Recognizing people's different responses to their environment, the variety of their work styles and the mixed range of their expectations and abilities is the key to achieving successful office installations. The aim must be to produce a degree of harmony between the individual's need for meaningful, satisfactory and creative work and the organization's demands for dedication and the application of skills and energy to achieve stated goals. Creating variety in people's work settings can encourage feelings that a personal contribution is being made by each and every person in the organization. Priority needs to be placed on:

- finding out above people's wants which can be conflicting and highly variable (see Strategy One);
- matching individual and group needs with appropriate work places (see Strategies Four and Five);
- thereby creating conditions for achievement.

Strategy Eight: Contribute to people's 'self-esteem' by focusing on their sense of well-being

Individuals in organizations see things differently from one another; a job that is satisfying and motivating for one person can be regarded as boring and pointless by another. Such discrepancies occur because human needs are ever changing; they can never be expressed as a fixed set of requirements.

In office design, lower-order needs (for safety and physical comfort) can normally be met by providing adequate spatial and environmental conditions which conform to appropriate building and health regulations. But there are reasons now why the focus of a designer's attention should be diverted towards people's higher-order needs.

Increasingly, people in organizations seek to experience a feeling of accomplishment and a sense of growth in the work they do. Unlike the meeting of lower-order needs, such higher-order needs cannot be satisfied by the provision of further and better environmental supports. Essentially, the desired outcome of self-esteem can be achieved only internally; it has to be given by each person to him or herself. The point here is that designers can consciously encourage such an outcome by demonstrating a paramount concern for the well-being of the people for whom they are designing.

CONCLUSIONS

The strategies outlined briefly here indicate that real opportunities for transforming organizational life occur where there is a recognition that solutions must be found which reflect people's increasingly complex and shifting needs. It is in the well managed organization, where people's tasks are generally rewarding and provide a sense of achievement, that the facilities management/design team can contribute most. By producing a greater sense of job satisfaction at the workplace, design can become an effective instrument of organizational success. It can make available human energy which would otherwise be dissipated.

In taking a strategic approach to workplace design, results need never be hit-or-miss; instead the design process becomes a precise tool for conveying an organization's essential nature. Clearly, the strategies outlined here must inevitably be modified to suit the specific needs of different organizations working in different parts of the world. But the overall purpose remains the same – to improve people's productivity by increasing the efficiency and pleasure of work.

SUMMARY

Facilities management entails the integration of people, technology and support services to achieve an organization's mission. It is concerned primarily with the quality of service to all stakeholders in the organization. In a service level culture the informed buyer assesses needs, agrees service levels and purchases services to meet them. The facilities organization must enable managers to focus on buying the best standard of service achievable within an agreed budget.

It is in well managed organizations, where people's tasks are generally rewarding and provide a sense of achievement, that the facilities management has most to contribute and can become an effective instrument of organizational success. Workplace design can transform organizational life. To help achieve this the organization needs to establish a single point of responsibility and concentrate its key management resources on the strategic role, which is to be an informed buyer of facilities services, including those of design. A facilities director or senior manager will be primarily responsible for tuning facilities to the needs of the business and in most organizations the role carries three related responsibilities for effective management of customers, assets and service.

Organizations also have to resolve issues concerning managing the facilities enterprise and privatization. A facilities management unit can help in this by evolving, consolidating and maturing, either as an autonomous business unit or as a separate independent company, that can deliver a guaranteed service to contract. Facilities, as property, also represent considerable value as the fixed assets of an organization and they need to be managed, repaired, maintained and replaced effectively to ensure their value is realized.

BIBLIOGRAPHY

Essential reading

Alexander, K. (forthcoming) *Developing Facilities for Competitive Advantage*, Butterworth-Heinneman, London.
Morgan, G. (1986) *Images of Organization*, Sage, Newbury Park.
Schonberger, R. (1986) *World Class Manufacturing: The Lessons of Simplicity Applied*, Free Press, New York.
Senge, P.M. (1992) *The Fifth Discipline: The Art and Practice of the Learning Organization*, Century Business, London.

Recommended reading

Davidow, W.H. and Malone, M.S. (1992) *The Virtual Corporation*, Harper Collins, London.
Garratt, R. (1994) *Learning Organisations*, Harper Collins Paperbacks, London.
Handy, C. (1989) *The Age of Unreason*, Business Books, London.
Mintzberg, H. (1994) *The Rise and Fall of Strategic Planning*, Prentice Hall, London.
Porter, M. (1985) *Competitive Advantage: Creating and Sustaining Superior Performance*, Free Press, New York.
Schwartz, P. (1991) *The Art of the Long View*, Doubleday, New York.

Further reading

Hampden-Turner, C. (1990) *Corporate Culture*, Hutchinson, London.

Handy, C. (1976) *Understanding Organizations*, Penguin Education, Harmondsworth.

Handy, C. (1994) *The Empty Raincoat*, Hutchinson, London.

Hayek, F.A. (1988) *The Fatal Conceit*, Routledge, London.

Heller, R. (1995) *Managing '95*, Sterling Publications, London.

Heller, R. (1990) *Culture Shock: The Office Revolution*, Hodder & Stoughton.

Kaufman, H. (1985) *Time, Chance and Organizations: Natural Selection in a Perilous Environment*, Chatham House, Chatham, NJ.

Mullins, L. (1993) *Management and Organizational Behaviour*, Pitman, London.

Peters, T. (1990) *Beyond Hierarchy*, Macmillan, London.

Schumacher, E.F. (1979) *Good Work*, Cape, London.

Semler, R. (1993) *Maverick*, Arrow Business Books, London.

3

Facilities management skills

OVERVIEW

Meaning is the message – and managing meaning will be one of the key organizational tasks of the next era. Alan Leigh

The way to successful facilities management relies on bringing to the operation of the organization an appropriate balance of business and technical skills. Facilities managers, as members of a professional discipline, have to function in a strategic role for this. The way in which their skills are used by organizations will determine the shape of the profession and its development. This is because it is a core management discipline and actively involves people, as facilities managers, in looking after the organization's mission, its business and its assets. It is these managers who can create the organizational vision and make the right choices to enable the organization to cope with complexity and the future. The successful facilities manager needs the power to effect change as well as the imagination to foresee its consequences. In the climate in which they will work in the future, their enabling skills will be needed to accommodate changes in both business needs and social practices.

If managerial ability and competence rather than technical capability is the basis by which they will be judged, they will, as 'hybrid managers', be required to understand the nature of problems and make things happen in the management of facilities. In many cases this could mean adding to a base of technical skills the necessary complement of business, social and personal skills. Not only should they have a vision of the service that will meet the requirements of business, but also the management skill to build and lead the facilities team to deliver to it. Technical professionals must learn a new language of function, performance and service and this will, in effect, become the language of 'design', 'specification' and 'contract' in facilities management. Similarly, management professionals must of course appreciate the technical environment in which they are functioning. Depending on the level at which they are operating, they will require all available tools and the ability to implement them. This particularly applies to the key skill – the ability to communicate.

New purchasing skills with the need for a strong purchasing capability and new contracting skills with the need for effective negotiating ability will be required by the informed buyer or intelligent client. Service managers must be contracted to ensure that the specified level of service is procured and delivered to the appropriate quality, on cost and on time. Effective feedback mechanisms are needed to ensure that the organization remains aware of the user's views at all times, and the service provider team must be organized and motivated to deliver quality services for all customers through well managed processes.

In the climate in which facilities managers will work in the future, their enabling skills will be needed to accommodate changes in business needs and social practices. To be effective he or she must have the authority and be prepared to accept full accountability for all actions. Radical personal and professional adjustment may well be required by many aspiring facilities managers. As successful managers they will not only need to manage well but will be viewed as a leader within the organization.

Keynote paper

Enabling strategies in planning for change
Craig Anderson

INTRODUCTION

Facilities management is about enabling organizational effectiveness. It is frequently referred to as providing support to business. Whilst the debate over precise definitions will labour on, the discipline should not be mistaken for a support function. Rather it is an enabling mechanism which responds to the evolving needs of business. Similarly, debates as to what it should include and how it should be structured will also continue, as organizations search for the optimum solution to meet their needs. It is these evolving business needs that are explored here. The following paper describes how facilities managers and organizations might respond to such needs. Above all it should be viewed against a backdrop of change – one of life's two certainties.

ORGANIZATIONS AND THE FUTURE

Organizations face a future full of uncertainties. For some, that future is some unforeseeable time ahead; for many, however, that time is now and the facilities manager will come of age in the role of a change-master. Possessing 20:20 vision in hindsight is easy. Looking back, for example, at the factors which contributed to the growth of the discipline of facilities management it is obvious how inevitable it was. Having a clear vision of things to come, however, is somewhat harder.

John Naisbit's book *Megatrends* (McDonald, London, 1984) produced ten predictions which would change the way in which we would go about our lives in the 1990s. To a greater or lesser extent, all ten predictions are based on objective facts of the way in which our world and those who share its passage are changing dramatically. But predicting the future is not so much the skill of a visionary but is the province of the enlightened.

Naisbit's latest book *Megatrends 2000* (with P. Aburdene, Morrow, New York, 1990) makes similar sweeping predictions which will shape the way in which the world will conduct its business around the new millennium. It is evident that only fools, and perhaps those experiencing an unsuccessful year in the futures markets, rush in to mitigate against the threats and take advantage of the opportunities afforded by these sweeping changes. The greatest fools of all, however, are those who choose to procrastinate when the window of opportunity presents itself.

Whilst the 'valour' vs 'discretion' choice is always likely to be a hard one when contemplating change, scant comfort can ever be gained by modelling the 'do nothing scenario'. Heraclitus (*c.* 500BC) stated that 'nothing endures but change'. The challenge for any management is making the right decisions at the optimum time. In the dynamic and competitive environment in which all businesses must operate, whilst managing change is a business success prerequisite, anticipating change is the real enabler. In the Lewis Carol classic *Alice in Wonderland*, Alice was faced with a choice between two roads; the one she took, as the Cheshire Cat said, depended upon where she wanted to get to. Unless we know where we want to be or what to expect when we get there, how can we plan for the future?

CHANGE DRIVERS

Hammer and Champy in *Re-engineering the Corporation: A Manifesto for Business Revolution* (Nicholas Brealey, London, 1993) summarized the three forces driving change within organizations as the three Cs – customers, competition and change. Globalization of business resulting from the removal of trade barriers, increased world democracy, improved telecommunications and emerging countries and technologies will result in more opportunity and greater prosperity across the world. The rise in small businesses and entrepreneurship also provides increased choice for consumers.

Increased customer, shareholder and stockholder expectations from products and suppliers, together with improvements in technology, result in increased competition which in turn places pressure on costs and service. Organizations, according to Michael Porter in *The Competitive Advantage of Nations* (Collier-Macmillan, London, 1989) may seek two kinds of competitive advantage: low cost or differentiation. Competitive advantage is a function of either:

● providing comparable buyer value more efficiently than competitors (low cost); or
● performing activities at comparable cost but in unique ways that create more buyer value than competitors and, hence, command a premium price (differentiation).

Such trends are forcing changes in the way businesses organize themselves for productivity and position themselves in the marketplace, as they seek to demonstrate the added value of their product or service.

PATTERNS OF CHANGE

Continuous pressure on cost and quality has forced the major restructuring of organizations across sectors and industries. Rapid increases in technology and automation invariably result in significant 'downsizing' of the labour force in the manufacturing sector, leaving only small teams of highly skilled 'knowledge workers'. In 1989, Chase and Garvin wrote of manufacturing organizations:

Who wins and who loses will be determined by how companies play, not simply by the product or process technologies that qualify them to compete. The manufacturers that thrive into the next generation, then, will compete, by bundling services with products, anticipating and responding to a truly comprehensive range of customer needs.

(Chase, R.B. and Garvin, D.A. (1989) 'The service factory', *Harvard Business Review*, August)

Many organizations, both product and service based, are recognizing that it is only through the provision of added-value services that they can overtake or distinguish themselves from the competition.

In the service sector, the IT explosion, whilst resulting in changing patterns of work, has perhaps failed to deliver the productivity gains and cost savings promised. Significant emphasis is being placed therefore on strategies that fall under a host of buzzwords such as 'restructuring', 'downsizing', 'rightsizing', 'delayering' and 'outsourcing' as they strive to reduce costs, improve service and quality as well as focus on their core business activities. The 1980s saw quality management high on most corporate agendas and total quality management has become the business philosophy.

The intelligent client, however, is still increasingly discerning, demanding more than assurances, in a continuous commitment to quality improvement throughout an organization. Business process re-engineering is the resultant discipline as organizations ask themselves not only 'what do we do?' and 'how can it be improved?' but more fundamental questions such as 'why do we do what we do at all?' In addition to looking inwards and re-engineering the organization, they are looking to their marketplace in order to develop strategic alliances with synergous or competitor organizations.

CHANGE ENABLERS AND ENTERPRISE CORES

Charles Handy wrote in 1989 of the emergence of 'shamrock organizations' (*The Age of Unreason*, Arrow, London). In essence these are organizations that have reorganized themselves to be flexible and responsive to market trends. Such 'leaner' and 'meaner' bodies operate with only a small core of professionals who work in customer-focused teams. Such project teams are supported by a flexible external and part-time labour force, the third leaf being contracted-out services and suppliers of non-essential services. Handy describes the fourth leaf as the flexibility and responsiveness dimension – the 'lucky' four-leaf clover providing competitive success.

The emphasis on a small core of essential people representing the knowledge base of the organization confirms Chase and Garvin's conclusions that, as fewer staff remain, they will be required to make a greater contribution in managing complexity and diversity, tracking quality and appreciating customers. Drucker has argued, as have others, that organizations in the future 'require knowledge works who will add value by thinking more like general managers, by appreciating a holistic view of the work process and suggesting innovative ways to enhance products and services'. (*Managing for the Future*, Butterworth-Heinemann, London, 1992). In the same way as quality improvements come about through the achievements of people, in the 'new organization' it is empowered workers who will be the catalyst for change. It was in 1990 that Peter Senge described the learning organization as 'a group of people who are continually enhancing their capacity to create their future' (*The Fifth Discipline: The Art and Practice of the Learning Organization*, Doubleday, New York).

PART-TIME AND CONTRACT WORKERS

Changing patterns of work resulting from technological innovations, economic pressures on productivity and changing social patterns resulting in more female workers entering or returning to the jobs market have contributed to the growth in part-time working. Organizations are now seeking to increase their ROI (return on investment) in staff by recognizing the need to provide opportunities to allow women and men to return to work whilst maintaining family commitments.

In 1989, Moss Kanter argued that: 'As contracting-out continues, and companies substitute supplier alliances for permanent employment, more people can expect to find their career in "producer service" industries' (*When Giants Learn to Dance*, Routledge, New York). The externalization of employment providing a 'just-in-time' workforce allows flexibility to organizations in hiring and firing temporary help to level imbalances in human resources and workload. The growth of 'contingent' rather than 'permanent' employment is a phenomenon synonymous with the contracting-out of services to third parties. This has accompanied the growth of facilities management as organizations seek to add value, improve quality and reduce the risks associated with service provision.

STRATEGIC SERVICE ALLIANCES

As the core of organizations diminish, and structures delayer management tiers, responsibility for the quality of service provision is increasingly being placed on the provider/supplier. The consequences of this are procurement strategies, which place increasing emphasis on the development of long-term relationships with suppliers. Such conflict-based strategies are getting less, as purchasers and providers realize the financial and other benefits associated with a relationship based on openness and trust. The benefits associated with small central cores of autonomous business units seeking to develop alliances with synergous organizations are increasingly being recognized. Supplier networks improve resource flexibility and provide added-value services to customers, whilst minimizing corporate overheads.

Networked companies operate on the principle that those activities which can be conducted and managed in a vertical way can be undertaken more cost-effectively by collections of specialist companies organized horizontally. This is a trend that is gaining momentum with those who profess to offer 'total facilities management' packages. Like 'partnership sourcing', such relationships have to be based on trust, with the alliances between companies in the network based upon ephemeral linkages that are part of the nature and duration of the shared venture.

PLANNING FOR CHANGE

The role of facilities management in organizations is to support the achievement of organizational goals. In order to respond proactively to enable organizational development, facilities managers must appreciate not only where the organization is going but act as a catalyst to the change process. As has been suggested elsewhere, the winners of tomorrow will look to the change process (chaos) as a source of marketing advantage, not as a problem to get around.

Traditional methods of planning for change are based on forecasts, and, despite errors from time to time, they are usually reasonably accurate. They are also based on the assumption that tomorrow's world will be much like today's – that they often work is because the world does not change. Eventually the forecasts will fail when they are needed most – in anticipating major shifts in the business environment. Planning for change involves identifying priorities in order that appropriate strategies might be developed. Drucker, in *Managing for the Future*, stated: 'In turbulent times, the first task of management is to make sure of the institution's capacity for survival, to make sure of its structural strength and soundness, of its capacity to survive a blow, to adapt to sudden change, and to avail itself of new opportunities.' The first priority in this scenario is survival, the second is achieving goals and the third is rational decision-making.

Using Drucker's priorities as a framework, how can facilities managers plan for and accommodate changes in business needs and social practices?

Priority One – Survival

In order to survive, organizations will require to remain competitive. Whilst profit maximization is not necessarily an agenda item for them all, others, including those in the public sector, will increasingly have to demonstrate value for money, as custodians and trustees of resources. Improvements in capital efficiency, whilst possible in the short term through 'downsizing' and 'outsourcing', are short-term fixes to more fundamental problems. Improvements in productivity and quality, and giving added-value at the same cost, will provide greater stability and organizational cohesion with reduced risks.

Improvements in productivity often result from investment in technology and facilities. However, the greatest return can often be gained from investment in people. Releasing the potential of the workforce sounds a bit like 'turning the asylum over to the inmates', but empowered workers – working either individually or in project teams sharing understanding and networking ideas – are the hallmark of progressive companies committed to total quality management.

Priority Two – Achieving goals

Organizations are essentially goal-seeking, and look to reduce discrepancies between where they perceive themselves to be and where they would like to be. It is essential that there is goal congruence between the evolving plans of business and the facilities enterprise. Facilities management operates most comprehensively in larger organizations, often with several hundreds or thousands of employees, where significant amounts of capital are often tied up in fixed assets. Such organizations can become large, heavily committed and often inflexible – almost like the dinosaurs in fact, which did not adjust well to sudden environmental changes as we are aware!

In an increasingly dynamic and turbulent business environment, facilities professionals will have to plan for change in the face of an uncertain future. As the future is uncertain, no single 'right' prediction or projection can be deduced from past behaviour. Acceptance of uncertainty, understanding its ramifications and making it part of reasoning involves the development of decision scenarios. Scenarios force people to think about the future in a more comprehensive and structured manner than traditional strategic planning. Most scenarios quantify alternative outcomes of obvious uncertainties – such as the price of a commodity in the future. But whilst such scenarios are of value, their main benefits come from involving staff and managers throughout the organization in the scenario planning process. Such involvement helps them to understand the changing business environment more intimately than they would have under the traditional planning process.

The role of scenario analysis exercises is in helping managers to structure uncertainty, as they are based both on a sound analysis of reality and on a change in the assumptions of decision-makers about how the world works. The key to future business success must involve developing alternative strategies shaped to meet alternative future states. Mintzberg and Waters (1985) describe such phenomena as 'intended' and 'realized' strategy. In essence the intended strategy is deliberate, realized as intended. Emergent strategy is that which results from patterns which are realized despite, or in the absence of, intentions. Deliberate and emergent strategies can be conceived as two ends of a continuum along which real-world strategies lie. Within the continuum, strategies include 'planned', 'entrepreneurial', 'ideological', 'umbrella', 'process', 'unconnected', 'consensus' and 'imposed'. In the new business world greater emphasis is to be placed upon those strategies at the 'emergent' end of the continuum.

Such approaches challenge our traditional attitudes that planning strategy involves a process of logical thinking and rational control. Whilst life is lived forwards it is understood backwards, and accordingly we make sense of the future from our experience of the past. Emergent strategy does not need to mean that management is out of control, only that in some cases it is flexible and responsive – in other words 'willing to learn'. It does involve, however, being prepared to adapt and respond to alternative courses of action. Such responsiveness comes from modelling uncertainty such as in the techniques of scenario planning.

Priority Three – Rational decision-making

As organizations seek to achieve desired ends they will adapt and modify their behaviour to changes in the environment. Those which are unable to do this will become extinct. The desire to be rational in decision-making is as much to do with the freedom and ability to allocate resources as it is with the opportunities available for error. Knowledge is all. Information changes the context of a message and provides the levels of confidence necessary for effective decision-making. Facilities managers are required to act as 'information brokers' and become 'big picture' oriented, as they assess the facilities management implications of evolving business plans.

THE FUTURE-ORIENTED FACILITIES MANAGER

Facilities managers in the future will practise their profession in a very different world from today's. The playing field may be familiar but the rules for winning the game will have changed dramatically. They will require to be flexible, skilled visionary leaders who can embrace change with a 'can do' mentality. Whilst technical and analytical skills will be required, it is the intuitive and managerial skills which can help managers to think holistically, creatively and flexibly, and to lead their customers and staff through seemingly chaotic environments. Successful facilities managers will recognize the importance of delivering high-quality services that meet the client's perceived expectations. They will develop interpersonal relationships and make decisions in conditions of ambiguity. As 'knowledge brokers' they can expect to be recognized as invaluable organizational assets.

Futurist Arthur C. Clark in his book *Profiles of the Future* has written that he 'does not try to describe the future, but to define the boundaries within which possible futures must lie'. Similarly the scenarios described above are not so much an attempt to predict the future as a framework for defining possibilities and expanding our vision of tomorrow. Clark wrote: 'If we regard the ages that stretch ahead of us as an unmapped and unexplored nation, what I am attempting to do is survey its frontiers and to get some idea of its extent – the detailed geography of the interior must remain unknown until we reach it.'

HOLDING A VISION

Uncertainty both about alternative future states and their results implies risk. However, uncertain futures, whilst not free from risk, need not result in negative consequences. Information provides the confidence in decision-making, just as organizational flexibility provides the ability to capitalize on new market opportunities. On the positive side, the emphasis on fewer managers doing more work implies that talented managers will be in great demand, managers who, in addition to being knowledge workers, will provide vision and leadership. In today's world, management means the ability to forecast, plan and control. In tomorrow's world of change, it will mean the ability to develop and communicate the vision and strategy in a language that everyone understands.

Facilities management in the future will be a team game. The team selected for each game will be according to the contribution individual players can make to the team performance. Whilst there may be more players on the substitutes bench, and more teams in the league, the regular team players will be those who possess talent, skills and fitness to compete. Talent may or may not be inherent, but skills can be acquired. Talented players in many cases aren't born – they learn. As Thomas Jefferson said: 'I'm a great believer in luck, and I find the harder I work, the more I have of it.' Fitness results from the ability to respond to new opportunities and to be flexible. Talent, skills and fitness, though, are only half the story. In true 'Roy of the Rovers' fashion, above all else, it is the desire to succeed – not just to compete, but to win. It is that vision, and its effective communication by the team manager, that carries the team to success.

SUMMARY

In the role of changemaster, the facilities manager should head a team which can compete as regular players in an organization. They should possess the talent, skills and fitness for this. As leaders with vision, and the ability to communicate effectively, he or she should be able to carry the team to success.

The skills required of a facilities manager are perhaps best described as follows. Organizations with a declared commitment to people, the workplace and the environment need individuals who can enhance their reputation through both attitude and action. A role of this kind, filled by such a person, would give the organization the scope to identify responsibilities necessary to plan, create and maintain an internal and external working environment conducive to the safe, effective and efficient carrying out of all business activities.

He or she would have the aim of setting the standard for facilities control across the business by providing proactive leadership for small highly trained teams together with a positive 'can do' approach. This would be mirrored by the team and it would manifest itself by working effectively through others, including the coordination of all contractor relationships. The ability to create an environment which matched the organization's open, non-hierarchical nature would also be achievable. It would give support to its belief that the right facilities result in a competitive advantage.

Are facilities managers being well enough fashioned to lead the facilities management scene in this way for the decade or so ahead? A strategic approach to facilities management, with the aim of producing a 'hybrid manager', requires the development of skills that are a balance between management and technical skills. This is not to underplay the importance of the technical competence required in the facilities team but this breadth of approach will not happen without clear vision, clear objectives and well developed quality systems. More work is needed on the human resource aspects of facilities management in order to get a deeper insight, without which these skills will be compromised and opportunities missed.

BIBLIOGRAPHY

Essential reading

Adair, J. (1986) *Effective Teambuilding*, Gower, Aldershot.

Alexander, K. (1993) *Managing Quality, Value and Risk: Essential Facilities Management Skills*, Centre for Facilities Management, University of Strathclyde.

Alexander, K. (1993) *Facilities Management Processes*, Centre for Facilities Management, University of Strathclyde.

Evenden, R. and Anderson, G. (1992) *Management Skills: Making the Most of People*, Addison-Wesley, Reading, MA.

Recommended reading

Adair, J. (1984) *The Skills of Leadership*, Gower, Aldershot.

Drucker, P. (1992) *Managing for the Future*, Butterworth-Heinemann, London.

Hammer, M. and Champy, J. (1993) *Re-engineering the Corporation: A Manifesto for Business Revolution*, Nicholas Brealey, London.

Porter, L.W. and Roberts, K.H. (1977) *Communication in Organisations*, Penguin, London.

Rees, W.D. (1988) *The Skills of Management*, Routledge, New York.

Senge, P. (1990) *The Fifth Discipline: The Art and Practice of the Learning Organization*, Doubleday, New York.

Further reading

De Bono, E. (1991) *Six Action Shoes*, Harper Collins, London.

De Bono, E. (1990) *Six Thinking Hats*, Penguin, Harmondsworth.

Buzan, A. (1993) *Use Your Head*, BBC Publications, London.

Chase, R.B. and Garvin, D.A. (1989) 'The service factory', *Harvard Business Review*, August.

Easterby-Smith, M., Thorpe, R. and Lowe, A. (1991) *Management Research: An Introduction*, Sage, London.

Handy, C. (1989) *The Age of Unreason*, Arrow, London.

Kanter, R.M. (1989) *When Giants Learn to Dance: Managing the Challenge of Strategy, Management and Careers in the 1990s*, Routledge, New York.

Mintzberg, H. and Waters, J. (1985) 'Of strategies, deliberate and emergent', *Strategic Management Journal*, Vol. 6, pp. 257–72.

Naisbit, J. (1984) *Megatrends: Ten New Directions Transforming Our Lives*, McDonald, London.

Naisbit, J. and Aburdene, P. (1990) *Megatrends 2000: Ten New Directions for the 1990s*, Morrow, New York.

Porter, M. (1989) *The Competitive Advantage of Nations*, Collier-Macmillan, London.

4

Professional practice

Professionalism – the ethical use of knowledge in the context of action.   Frank Duffy

Facilities management practice takes a positive look forward both at what the future holds and at possible steps in changing the ways in which services are delivered. It provides an understanding of the legal, procedural and contractual setting, and is a framework which recognizes alternative perspectives – FM as a discipline, FM as a profession and FM as a market.

Facilities managers fulfil the role of intelligent client, defining requirements, negotiating service levels, purchasing services and monitoring contract performance. They possess a mix of business, management and technical knowledge and skills, and are ready to assume broad responsibilities at an increasingly senior level in organizations. Professional associations have been formed to represent their interests and promote facilities management in many parts of the world, often collaborating as networks.

Facilities management consultants offer a range of professional services in design and specification, sourcing and contract management. A number of service organizations act as managing agents and management contractors to provide and deliver professional services and integrated, often comprehensive total facilities management (TFM), packages of services to organizations with clearly identified needs.

Consultants and contractors both seek to operate in the non-core activities of organizations in both the public and private sector. Services are bundled into packages and offered under the title 'facilities management'. Estimates of the size of the market vary widely – from £12 bn to £47.5 bn for example – depending on the definition and scope of the services included, but all assume that contracting-out and facilities management are synonymous. The grouping of site services and packaging of service contracts is *not* facilities management. On the whole these are operational services provided to rigidly specified terms and conditions, with little scope for innovation. Assessing the market potential of facilities management is not simply a matter of collecting together information about the extent to which independent contract services are contracted out.

Reviews, such as those published by the Centre for Facilities Management in *Facilities Management 1994* and *1995*, of the services presently offered by the market show that they fall well short of the requirements being set by 'intelligent clients'. The contract services can too easily be rooted in the core businesses of the host company – construction, maintenance, cleaning, security and catering – and need to look beyond packaging and aim for value-added services. Professional bodies in the construction, hospitality and information technology industries now recognize the opportunities that facilities management provides to members to develop new, value-added services, better integrated with an organization's business needs.

Meanwhile, from management buy-outs and spin-offs, new service organizations can have a clearer understanding of a client organization's requirements, and previous experience helps in organizing, designing and specifying the service required. The facilities management 'industry' – major clients, professional associations, contractors – needs to pull together to create forms of partnership, codes of practice and standardized contracting practices if the principles of facilities management are not to be compromised.

Keynote paper

A sea-change in facilities management
Edward Finch

INTRODUCTION

The introduction of facilities management as the 'practice of coordinating the physical workplace with the people and the work of the organization' understates its significance as a discipline. Facilities management is far more than a set of tasks; it deals with facilities in an entirely new way. The paradigm here concerns the link between organizations and buildings and the way this is established and sustained.

PROPERTY AS A FACILITY

The crisis that has prevailed in recent decades in almost all property sectors has, in part, helped fuel the rapid evolution of facilities management. Historically a good building location ensured a reliable investment for the property owner. Other building deficiencies such as environmental inadequacy, poor performance, inflexibility and costly maintenance were often obscured by this forgiving characteristic. Advanced telecommunication systems have dramatically affected this situation.

Information technology has removed many constraints surrounding the possible location of organizational activities. As a result, attention now focuses on the quality and suitability of the building as a product and, in particular, its capacity to accommodate information technology.

Other trends have served to reinforce this attentiveness. Knowledge workers, who have become the mainstay of employees in the office sector, make increasing demands of the workplace. Not only must the building be supportive of work, it should provide a key motivational element in the work environment. It was belated recognition of these trends that resulted in the property crisis of the 1970s and 1980s. Property owners had 'rested on their laurels' only to find that their assets had become prematurely obsolete. Buildings were unable to satisfy these demands because of many intractable flaws such as inadequate space provision for services. These failures were highlighted in the ORBIT report produced by Davis *et al.* in the USA in 1985.

The general outcome of these events is that property managers can no longer assume that properties with the right location will sustain their value. Only by continued monitoring of and adjustment in response to the changing technological and business environment can property value be assured.

MANAGER OF CHANGE

Against this backdrop it is clear that a permanent management structure for coping with change in the built environment is required. The job of facilities manager represents such a 'manager of change', being concerned with preventing the demise of a facility resulting from functional or technological obsolescence. This contrasts directly with the

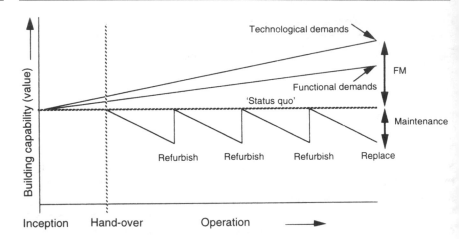

Figure 4.1 The distinction between maintenance management and facilities management

role of the maintenance manager, who attempts to combat the effect of physical deterioration, the objective being to maintain the status quo of the building. This distinction between maintenance management and facilities management is illustrated in the conceptual graph of building capability versus time shown in Figure 4.1.

Periodic maintenance allows the building capability to be recovered to a level comparable with the building's initial condition. However, a gap continues to emerge throughout the building's life because user expectations increase and, indeed, change in nature. This may result from external innovations in information technology imposing greater demands on the building, or internal organizational changes requiring a readjustment of functional capability. One important feature shown in Figure 4.1 is that the maintenance gap only arises once the building is in use, whereas the technological and functional gaps can arise from the point of inception, the briefing stage. This is in part an unavoidable result of the lag time between inception and completion, during which time changes may arise. The facilities manager does, however, have a key contribution to make at the inception stage ensuring, among other things, that knowledge relating to changes of previous buildings is considered in the new building.

It is no longer appropriate to consider the building process as a 'one-shot' process. This approach was described by J.C. Vischer in *Environmental Quality in Offices* (Van Nostrand Reinhold, New York, 1989) as the analytic approach whereby, provided sufficient attention is given to the briefing stage, it is possible to get it right the first time. The contrasting synthetic approach, also described by Vischer, applies an evolutionary model to the building. This requires an ongoing system of monitoring and response in relation to the building users. The onus thus falls on the facilities manager to continually bridge this gap.

MOTIVATING THE KNOWLEDGE WORKER

Buildings not only facilitate work, they can actually enhance it. Building operators should not simply be attempting to fulfil some minimum constraint but should be continually striving to maximize the building user's potential. Ironically it is the increasing cost of personnel rather than buildings that has focused our attention on building performance. In *Behavioral Issues in Office Design* (Van Nostrand Reinhold, New York, 1986) Wineman calculated that, in monetary terms over the 40-year lifecycle of an office building, 2–3% is generally spent on the initial cost of the building and equipment, 6–8% on maintenance and replacement and a staggering 90–92% on personnel salaries and benefits. Travel time between functional areas, levels of disruption and accessibility of information are all examples of work-related problems that can be improved by careful design and management.

Attention in facilities management has shifted from alleviating job dissatisfaction to fulfilling job satisfaction. As observed by Herzberg (in Nord, W.R. (ed.), *Concepts and Controversy in Organization Behavior*, Goodyear, Santa Monica, CA, 1976) these are not extremes of a single continuum: 'The opposite of job satisfaction is not job dissatisfaction but no job satisfaction; and similarly, the opposite of job dissatisfaction is not job satisfaction but no job dissatisfaction.' Environmental conditions which influence job dissatisfaction, such as 'hygiene factors', have until recently been the main preoccupation of building operators. However, there is a growing awareness that 'motivators' such as achievement, recognition and the work itself (which determine job satisfaction) are fundamentally influenced by the built environment.

The evolution in building users' expectations is shown in Figure 4.2, which is based on Maslow's triangle described in 'A theory of human motivation' (*Psychological Review*, Vol. 50, 1943). This shows the transition which has occurred from the expectations of the office worker of the middle twentieth century to the knowledge worker of today. Maslow's argument was that, once lower-level needs such as physiological and safety needs are met, then, at once, other higher needs emerge. In the context of the built environment, once environmental conditions such as heat and light are satisfied, the individual becomes dominated by the unsatisfied needs and environmental conditions cease to be important to the individual's current dynamics.

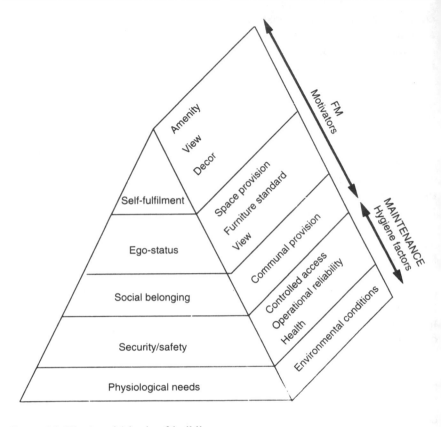

Figure 4.2 The 'needs' basis of building users

The belief that a happy worker is a productive worker can, however, be misleading. E.H. Lawler and L.W.L. Porter investigated the relationship between satisfaction and productivity and found that the correlation was very weak ('The effects of job performance on job satisfaction', *Industrial Relations*, Vol. 7, 1967). This contrasted with the correlation between satisfaction and two other factors, absenteeism and turnover, which were shown to be strong. They proposed a modified causal link, in which performance (productivity) resulted in rewards which in turn resulted in satisfaction. In other words, the cause–effect is in the reverse direction. The extent to which rewards led to satisfaction for individuals was moderated by their expectations. Clearly, a facilities manager in this context can be seen as a provider of reward – such as additional work space, improved working conditions and better views, etc. Lawler and Porter also suggested that the key to this match appears to be perceived equitable distribution of rewards.

Questions raised in relation to facilities management by these findings include the following.

- Should provision of resources such as space and specialized amenity be related to reward or should it be entirely based on need?
- Should reward be allowed to lead to privilege within an organization's building, or should it be viewed solely as a collective incentive to attract and retain high calibre employees?
- Can information reporting employee satisfaction in relation to building performance be used as a significant performance indicator, or should the facilities manager seek a more sophisticated measure of building effectiveness?

CONTRACTING METHOD

Innovations in facilities management contracting have closely paralleled those of the construction industry introduced a decade earlier. The internal pricing of transactions has prompted the consideration of buying the same service externally. Reinforcing this process has been a mandatory requirement for market testing, exemplified in many UK public sector organizations. This may include not only the operational services (cleaning, security, etc.) but also the management service – the core strategic element of facilities management. In construction the traditional method of procurement involves the use of a general contractor, who carries a degree of overheads in terms of physical assets and non-management staff, together with management staff. The analogous situation in facilities management is the fully outsourced package. Subcontractors are employed by the facilities management contractor for specialized services, although many operational activities are undertaken by the main contractor.

Management oriented procurement systems have become much more common in the construction industry over the last two decades. As explained by J.W. Masterman in *An Introduction to Building Procurement Systems* (E & FN Spon, London, 1992), this has been in response to the growing prominence of the subcontractor and the growth in size and complexity of buildings and building techniques. This trend is also evident in facilities management. A management oriented system involves the payment of a facilities management organization on a fee basis to administer the management of the facility. Subcontractors may be employed either by the client or the facilities management group, but the main facilities management group possess only management staff and do not carry any overheads in terms of capital assets or operational staff.

PROFESSIONAL PRACTICE

Table 4.1 Generic forms of facilities management contracting

Contract type:	In-house	Outsourced – main contractor	Outsourced – management based
Pricing system:	Internal transfer price	Fixed price or profit-sharing (partnering)	Fee based
Involvement of client:	Negotiation	Single-shot specification	Negotiation
Required expertise of client:	High	High	Low but informed

Table 4.1 detailing contracting methods has many variants within it in terms of risk apportionment, contractual relationships with subcontractors, incentives, profit-sharing, etc. No single method provides a panacea for all client organizations. Their characteristics in terms of cost, quality, risk and client participation depend largely on the nature of asset specificity described later in this chapter. The issues which need to be considered in identifying suitable partners in a facilities management contract, whether it be in-house or outsourced, form part of what is known as the transfer pricing system and are detailed in the following section.

TRANSFER PRICING SYSTEMS

What is a fair price to pay for a workplace? How can costs for an intangible like 'environmental quality' be justified? Large differentiated organizations, with diverse and complex business activities, continually face the problem of identifying ineffective operations. Under such scrutiny, facilities management must take up a defensible position.

Failure to deal with hidden inefficiencies has damaged large corporate giants at the hands of smaller organizations which are capable of identifying inefficiencies. In response to this, the concept of profit centres has emerged. This requires individual operations in an organization to apply profit as a constraint or objective. (Various authors, including Lawrence and Lorsch in *Organization and Environment: Managing Differentiation and Integration* (1967), argue that profit is rarely an objective in an organization, but profit must be realized in order to sustain an organization and as such is more accurately described as a constraint.) The use of profit centres also implies a greater degree of autonomy and decision-making at the local level.

Given the degree of interdependence which exists between the various operations in an organization, the definition of profit requires examination. This leads to the concept of transfer pricing, which is concerned with the pricing of services which are offered by various operations in an organization. From the perspective of facilities management, this has a twofold significance. If facilities management is treated as a discrete operation, the unit must be capable of yielding a tangible profit as a profit centre. Alternatively, if facilities management is treated as a supportive organizational function, the question of the amount of rental paid by profit centres becomes a key consideration (this essentially involves the question of transfer pricing). Whichever approach is used, the cost accountability of facilities management has become more prominent than ever before.

Transfer pricing deals with more than the transfer price: it addresses the whole mechanism of coordinating transactions, both external and internal. Williamson, in *The Economic Institution of Capitalism: Firms, Markets, Relational Contracting* (Free Press, New York, 1985), identifies the three stages of a transaction as:

- contact stage, when the parties look for suitable partners;
- contract stage, when the parties negotiate the contract terms;
- control stage, when the contract is completed in accordance with the terms identified in the contract stage.

A transfer pricing system comprises the following elements (modified from van der Meer-Kooistra's article in *Management Accounting Research* (No. 5, 1994, pp. 123–52):

- definition of the application area to which the rules apply (e.g. cleaning, security, etc.);
- desired characteristics of the transaction (e.g. business-economics based relation, which can be readily set up and administered);
- extent of authority of profit centres to determine the price and the supplier or buyer;
- basis for determining the transfer price (e.g. cost definition or market price definition with internal contracts);
- basis for adjusting the transfer price (when are changes allowed and who has to be consulted);
- transaction terms (service specification, quality standards and quality control);
- means of contracting (outsourcing, management contracting, partnering, in-house);
- arbitration process;
- consultation process (with regard to the drafting, functioning and adjustment of rules);

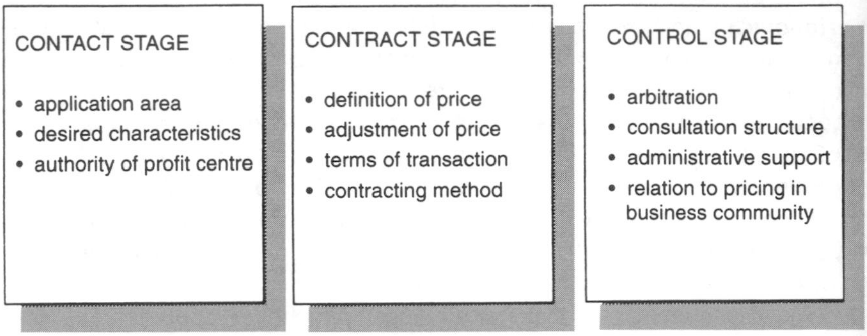

CONTACT STAGE	CONTRACT STAGE	CONTROL STAGE
• application area • desired characteristics • authority of profit centre	• definition of price • adjustment of price • terms of transaction • contracting method	• arbitration • consultation structure • administrative support • relation to pricing in business community

Figure 4.3 Elements of a transfer pricing system

- administrative support (including IT systems to record and process transactions);
- relation to fiscal transfer pricing system (i.e. the business community at large, benchmarking being an important facet).

The relationship between the three stages of a transfer pricing system and each of the elements of a transfer pricing system listed above are shown in Figure 4.3.

ASSET SPECIFICITY

Again according to Williamson (quoted above) the nature of transactions can be explained to some extent by asset specificity which describes the extent to which a given product or service is tailored to a specific customer or customer type. The greater the degree of specificity, the more important it is to establish ex-ante and ex-post protection against opportunistic behaviour by one of the parties. Facilities management is characterized by varying levels of asset specificity depending on the complexity of the building(s) and its (their) location, among other factors. The four key types of asset specificity relevant to facilities management are as follows.

- Site specificity represents the advantages derived from the proximity of partners (e.g. a security organization close to a facility).
- Physical asset specificity occurs when investments in tangible assets have been undertaken specifically for the other party's requirements. (In the most extreme case this is exemplified by bespoke buildings or refurbishments.)
- Human asset specificity arises from the experience gained from doing

a job, which inevitably increases knowledge about special activities (e.g. performance characteristics of a heating system).

- Dedicated assets are large investments in capacity (e.g. business park developments) which, although they are not bespoke, present problems in finding another buyer who wants to use the spare capacity.

An understanding of the extent and nature of asset specificity is critical to the formulation of an appropriate contracting method in facilities management. The provider of the facilities management service and the user (client or facilities manager) are both vulnerable to opportunistic behaviour by the other if issues of asset specificity are not considered. A common example is the inevitable process of human asset specificity which arises from the incumbent operators of a facility who undergo a considerable learning curve during the duration of a project. As a result they are in a position to demand a premium price when the contract is renewed.

Another illustration of the influence of asset specificity is the conflict which inevitably arises between the facilities manager acting as landlord and simultaneously acting as business facilitator. A refurbishment tailored for a specific organizational unit may be attractive from a functional point of view but may not be justifiable in terms of the contribution to the value of the built asset.

THE FACILITIES MANAGER AS INFORMATION PROVIDER

One of the difficulties of contracting out facilities management is the disappearance of key individuals who can provide advice at a strategic level within an organization. The facilities manager, in relation to property management, is in a unique position to provide strategic information. Often, however, even when facilities management is in-house, the quality of information required by senior executives is all too often lacking. With a financial based reporting system, buildings are invariably perceived as a liability which has to be tolerated. Recent evidence, however, suggests that senior executives are seeking information across an entire range of critical factors, the most commonly cited (in addition to cost structure) being:

- product quality and innovation;
- customer satisfaction;
- management development;
- change management and development.

The KPMG 1990 survey of leading companies highlighted the paucity of information provided to senior executives, which tended to be biased

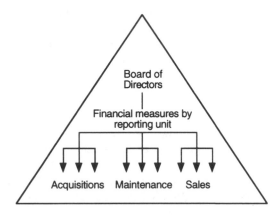

Figure 4.4 Traditional form of property reporting structure

towards financial indicators and concentrated on internal self-determined parameters. This results in a narrow view of the business with insufficient information on non-financial and external factors.

Figure 4.4 shows the traditional form of property information provided to senior executives. This contrasts with the strategic information shown in Figure 4.5 which should be used to communicate with top management.

WORKPLACE LEGISLATION

One of the arguments for outsourcing facilities management is the degree of specialism that is now involved. This is illustrated by the burgeoning of environmental, health and safety legislation in the US and Europe. it is becoming increasingly untenable to provide in-house staff with sufficient expertise in all areas of legislation. The EU Directive on 'the minimum safety and health requirements for work with display screen equipment' is one case in point. In the US, recent legislation called the Americans with Disability Act (ADA), which went into effect in July 1992, is expected to have an equally profound effect on US facilities management.

Employers, in compliance with the above EU Directive are now required to 'take appropriate measures to remedy the risks to which VDU operators are exposed'. Any employee who 'habitually uses display screen equipment as a significant part of his normal work' falls into this category. The Directive applies to the entire workstation, including

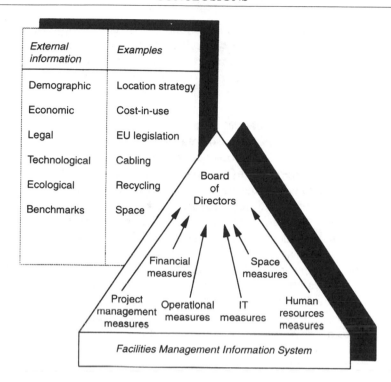

External information	Examples
Demographic	Location strategy
Economic	Cost-in-use
Legal	EU legislation
Technological	Cabling
Ecological	Recycling
Benchmarks	Space

Board of Directors

Financial measures

Space measures

Project management measures

Operational measures

IT measures

Human resources measures

Facilities Management Information System

Figure 4.5 Modern form of property reporting structure

display screen, keyboard, peripherals, document holder, work chair, work desk and the immediate work environment'.

The Directive also imposes requirements concerning environmental conditions such as lighting, noise levels, heat, humidity and radiation. Lighting schemes must not produce glare and adjustable secondary lighting must be available. Equipment heat must not cause discomfort and an adequate level of humidity must be maintained. This clearly raises fundamental questions about the capacity of building operators – and indeed the enforcers of the legislation – to measure these variables. What is clear from the detailed nature of this kind of legislation is that a considerable level of technical and legal know-how will be required for managing the workplace of the future.

CONCLUSIONS

Several facilities management issues are emerging from the present scene and will continue to evolve in the forthcoming decade. For example, it is likely that the most effective FM contracting policy will continue to be the one which considers the interests of the FM employees in terms of job security, opportunity and reward. The clearer the career ladder which exists in outsourced FM organizations the greater the stimulus and creative challenge. However, job security needs equal consideration.

The modelling of occupancy patterns is another issue. Occupancy patterns are becoming far less predictable with the demise of the 9-to-5 job and the advent of telecommuting. An article by Becker in *Facilities Design and Management* (Vol. 10, No. 2, 1991, pp. 48–51) makes that clear. Innovative solutions need to be introduced to monitor occupant movement and to resolve peaks and troughs in space demand. Automatic identification, radio frequency tracking, machine vision and other IT devices may assist in collecting and interpreting such information. Finally there is the provision of stimulating as well as comfortable environments. This deals not only with the setting of thermal and lighting conditions but variation of these parameters, spatially and over time. The realization that human activity has a cyclical pattern suggests that a fixed environmental condition is undesirable. Instead, an environment which is emphatic to this cyclical pattern needs to be considered.

SUMMARY

In each sector of the economy, organizations operate quality managed facilities using well developed quality systems supported by effective information management systems. A balanced and cohesive team of in-house managers, supported by consultants, are led by professionally qualified facilities managers. Regular audits provide feedback on quality, value and risk, and effective quality systems include regular training with programmes of organizational learning. Organizations have longer-term relationships with carefully selected service organizations giving value-added services. Demand has to be the impetus for this scenario of the professional practice of facilities management.

In any review of facilities management practice, forms of contracting are likely to come under scrutiny first. Purchasing and contracting practice which is at variance with the principles of facilities management is one suspect area. Long-term objectives can be compromised by short-term contracts, and defensive, adversarial relationships can stem from inappropriate terms and conditions. Any opportunities for rationalization and improvement through integration will be limited while there is a preference for short-term, separate subcontracts. Joint venture and partnering arrangements are more appropriate forms of relationship to meet facilities management objectives.

To sustain demand-led activity, and to deliver a promise of improved quality with controlled costs, requires an 'intelligent client'. Smaller organizations can form consortia to improve their skill base and purchasing power. A structured education and training provision to develop facilities management skills and offer supporting advisory services can promote the necessary development of quality systems, and ensure effective processes for marketing, design, selection and delivery.

BIBLIOGRAPHY

Essential reading

Argyris, C.S. and Schon, D.A. (1984) *Theory in Practice: Increasing Professional Effectiveness*, Jossey-Bass, San Francisco.
Becker, F. (1990) *The Total Workplace: Facilities Management and the Elastic Organization*, Van Nostrand Reinhold, New York.
Schon, D.A. (1983) *The Reflective Practitioner: How Professionals Think in Action*, Basic Books, New York.
Senge, P.M. (1992) *The Fifth Discipline: The Art and Practice of the Learning Organisation*, Century Business, London.

Recommended reading

Davis, G.F., Becker, F., Duffy, F. and Sims, W. (1985) *ORBIT-2: Organisations, Buildings and Information Technology*, The Harbinger, Newark, CT.
Editorial (1990) *Practice*, in Facilities Management International Conference Proceedings, Centre for Facilities Management, University of Strathclyde.
Editorial (1990) *Professional Development*, in Facilities Management International Conference Proceedings, Centre for Facilities Management, University of Strathclyde.
Lawrence, P.R. and Lorsch, J.W. (1967) *Organization and Environment: Managing Differentiation and Integration*, Irwin, Homewood, IL.
Maren, M. (1991) 'Productivity palace: the surprising science of workplace effectiveness', *Success*, Vol. 38, No. 7, September.
Maslow, A.H. (1943) 'A theory of human motivation', *Psychological Review*, Vol. 50.

Further reading

Becker, F. (1992) 'Excuse me: I think that's my desk', *Facilities Design and Management*, Vol. 10, No. 2, pp. 48–51.

Herzberg, F. (1976) 'One more time: how do you motivate employees?', in Nord, W.R. (ed.), *Concepts and Controversy in Organization Behavior*, Goodyear, Santa Monica, CA.

KPMG Management Consultants (1990) *A Survey of Leading Companies 1990 – Information for Strategic Management*, KPMG, London.

Van der Meer-Kooistra, J. (1994) 'The coordination of internal transactions: the functioning of transfer pricing systems in the organisational context', *Management Accounting Research*, No. 5, pp. 23–52.

Preiser, W.F.G. (ed.) (1993) *Professional Practice in Facility Programming*, Van Nostrand Reinhold, New York.

Schwartz, P. (1991) *The Art of the Long View*, Doubleday, New York.

Steele, F.I. (1986) *Making and Managing High Quality Workplaces*, Teachers College Press, New York.

Taylor, B. and Graham, C. (1992) 'Information for strategic management', *Management Accounting*, January, pp. 52–4.

Vischer, J.C. (1989) *Environmental Quality in Offices*, Van Nostrand Reinhold, New York.

Williamson, O.E. (1985) *The Economic Institutions of Capitalism: Firms, Markets, Relational Contracting*, Free Press, New York.

Wineman, J.D. (1986) *Behavioral Issues in Office Design*, Van Nostrand Reinhold, New York.

Quality management

5

OVERVIEW

Facilities management is a total quality approach to sustaining an operational environment and providing support services to meet the strategic needs of an organization.

Centre for Facilities Management

Techniques for managing quality, value and risk are key skills for facilities management. Processes for developing and managing social and service partnerships create the conditions for continuous quality improvement and, as a result, managing customer expectations and meeting their requirements implies a total quality approach. Facilities management is this kind of approach applied to developing and operating buildings and delivering support services which contribute to achieving business objectives.

The quality managed facilities (QMF) framework, as developed by the Centre for Facilities Management, can be described as a quality matrix for managing customers, service and assets at a strategic, tactical and operational level. It identifies the processes involved and proposes a course of action for facilities management. QMF can ensure that facilities conform to an organization's requirements, add value by contributing to achieving objectives and reduce the risks of business. It emphasizes the importance of an organization's quality culture to the effective management of facilities. The full costs of quality must be recognized. Total quality management recognizes that these costs can be reduced with appropriate attitudes, improved processes and effective teamwork, all geared towards the customer.

Quality managed facilities investigates management processes and relates facilities issues to total quality and customer care. Given that the only purpose of any business is to create a customer suggests a number of steps to be identified on the quality journey to systems capable of audit to international quality standards. The benefits of quality initiatives will take time to emerge.

The concepts of total quality management need to be considered as part of any cultural changes necessary within organizations. Total quality issues touch on the quality achieved in a range of management systems including management (policies, planning, organizational), people (motivation, training, utilization), operations (etablishing, maintaining, enhancing) and information (collecting, storing, disseminating and using).

Techniques of evolving a quality manual, and of auditing and monitoring quality, need to be considered together with management of change. Customer focused organization and aspects of product, service, information and environmental quality also need to be considered, as do the development and understanding of the use and appropriateness of quality management systems.

The 'added value' to be gained through a total quality approach is being increasingly recognized in business and so is its scope of application. TQM focuses attention on all activities and processes in terms of satisfying the requirements of customers. Everyone in an organization has 'customers' who rely on the quality of their performance. For an organization operating 'quality circles' the key to success is for teams within the organization to function and produce results as quickly as possible. TQM can then sustain the working environment and provide support services to meet the strategic needs of an organization.

Keynote paper

Facilities quality management
John Grigg

INTRODUCTION

Quality in facilities management should permeate every part of an organization and result in substantial benefits, both internally and externally. Total quality management (TQM), as the comprehensive application of quality to the facilities field is called, is not a technique or a campaign but a way of business life, and one which facilities managers cannot afford to ignore. This paper sets out the principle of TQM, relates it to what has to be done in the facilities management field, and then sets out how to do it and who should be involved.

Many organizations apply the concept of total quality management in one way or another and it is known by different titles in various organizations, e.g. TQM, quality service, customer care, quality improvement programme, and so on. The term TQM embraces all quality-led initiatives and methods of working.

TQM DEFINED

The power of total commitment to TQM is easily demonstrated by the rise of Japanese industries in the major world markets since the early 1950s. To attribute all Japanese success to TQM is of course an oversimplification. Most management gurus, however, and the Japanese themselves, acknowledge that the cornerstone of their postwar success has much to do with the adoption and total belief in TQM – the obsession with giving the customers what they want and continually striving to better current performance.

A good example of the power of TQM within an industry sector is the recent history of the world-wide semiconductor market. The Japanese were relatively late to adapt TQM in their semiconductor market, but when they did the results were impressive. In 1975 three of the top ten semiconductor companies were Japanese and these were all in the bottom half. By 1986 the number had increased to five, with positions one, two and three all occupied by Japanese companies. In 1988, six Japanese companies featured in the top ten and they continued to dominate the top three. All this was at the expense of American and European manufacturers.

Examples like this are typically referred to as the 'Japanese miracle'. But, of course, there is no miracle, just a commitment to hard work and an active desire to do better at all levels of the organization. Traditional management thinking has to change in order for TQM to be effective and to achieve sustainable benefits. First of all there has to be top-level commitment to the concept, not just lip-service to a new campaign. Secondly, management's role has to change from 'rule-maker' to 'facilitator'.

A useful phrase illustrates the revised role – that is 'removing the boulders from the runway'. It is essentially an enabling process. This means that the usual reporting pyramid, showing board members and senior executives at the top with supervisors and workers at the broader bottom, is inverted as shown in Figure 5.1. This gives a truer picture of

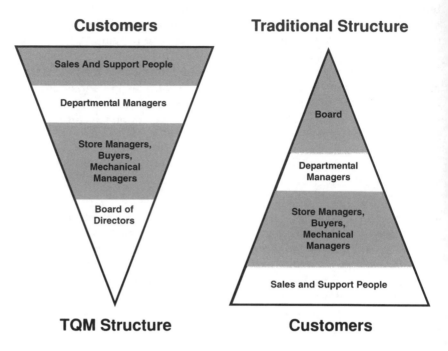

Figure 5.1 The necessary structural change in management style

the senior managers within an organization supporting the operatives, enabling them to carry out their work, and with the customers at the top.

So what is TQM? Dr W. Edwards Deming and the British Deming Association have contributed more than anyone to developing the modern approach to TQM and have helped shape the following definition.

- Quality is continually satisfying customer requirements.
- Total quality is continually satisfying customer requirements at least cost.
- Total quality management is meeting these requirements at least cost, through harnessing everyone's commitment.

So what is quality? Deming states: 'Good quality does not necessarily mean high quality. It means a predictable degree of uniformity and dependability at low cost with a quality suited to the market.' For instance, the perceived quality of a Rolls-Royce is obviously different to that of a VW Beetle. However, both consistently suit their customers' requirements and can therefore be said to be meeting a quality demand, as Figure 5.2 shows.

VW Beetle	Rolls Royce
Inexpensive	Space
Reliable	**Air-conditioning**
Always starts	Internal luxury
Runs forever	**Electrical seat adjustments**
Easy to maintain	Prestige
Durable	**Smooth, quiet ride**
Easy to park	

Figure 5.2 What is quality? The Rolls-Royce vs. Beetle comparison

TQM AS APPLIED TO FACILITIES MANAGEMENT

Facilities management is essentially a service to the core business of an organization. Given the more usual functions under most FM departments' responsibility such as cleaning, building maintenance, space planning, security, catering, engineering and so on, it is easy to understand that service to an internal or external customer is the *raison d'être*. In this area relevant to the FM practitioner Deming raises and answers a common issue.

> In the service industry, most workers do not think they have a product. They think they just have a job. They do have a product ... service.

Satisfying customer requirements must be the main goal of every facilities manager. Customers expect quality because they want a service that is reliable, they want it to meet their requirements and they want value for money. It is not a 'luxury', it is not meeting internal standards, it is agreeing what the standards are with the customers and continually achieving them by encouraging the commitment of all.

Quality in facilities management means satisfying customer requirements, reducing costs by getting things right first time and avoiding waste by eliminating errors. It does not, however, mean doing things on the cheap.

PRINCIPLES OF TQM

The principles behind TQM which help achieve customers' requirements are contained in six basic principles. They explain how TQM can be implemented and the 'rules' by which it is operated:

- The philosophy – prevention not detection.
- The approach – management supported.
- The scale – everyone responsible.
- The measure – the cost of quality.
- The standards – right first time.
- The theme – continuous improvement.

Using these six basic principles the facilities manager can ensure that the requirements of the customer are met at all times. Closer examination of each principle reveals how the facilities manager can apply them to help achieve a quality service.

The philosophy – prevention not detection

It sounds so simple. In fact everything to do with TQM is just that. However, there are many examples where detection has been the norm rather than preventing the error in the first place. Examples include things like the specification for a suspended ceiling for a fixed tile – if all the mechanical plant is installed above the suspended ceiling, it makes access for maintenance an extremely difficult and costly operation. Or take the siting of the waste bins in a kitchen area. If people have to carry used tea bags or throw them across the kitchen, the implications are obvious: stained floors and walls, disgruntled customers and shoddy work areas. Although these examples are simple, it is not difficult to think of many more that could have been operated more efficiently with some forethought.

The approach – management supported

The crucial word here is 'supported'. If the facilities team are aware that a TQM service is something they have to do rather than genuinely want to do, it will make implementation much more difficult. The facilities manager has to set the rest of the team an example of how to operate TQM. The manager should be seen to support and motivate his teams by open communication and total commitment. Quality service has to be seen as a way of life, not a campaign. The facilities manager needs to convince his people that it is the norm to question procedures, instigate change and talk to the customer. It needs high-level management support so that the whole team can see TQM is not just something imposed on them. They have to be convinced that the senior management have embraced and understood what TQM means. The facilities manager should help, train and support his team to make their own decisions in their own areas. The team should be looked on as the front line in communication with the customer. It doesn't matter how good the facilities manager is, if the front-line team are not communicating well with the customer, then a quality service cannot be offered. It is the process of 'removing the boulders from the runway'.

The scale – everyone is responsible

Everyone has to be responsible for delivering TQM. Even the people who may not have a direct contact with external customers should realize that their internal customers are just as important. The facilities team need to look at the work they do and ask themselves the following questions: Is the job I am doing going to help my customer?

Does it achieve the standard I have agreed with my customer? Is it necessary or am I doing it because it has always been done this way? How can I improve the job to increase the level of service I give to my customer without increasing the time taken or the costs? Is there someone better able to tackle this task, giving them more responsibility and freeing up the facilities manager's time? The recurring question should be: Am I taking responsibility for improving the quality of my work?

The facilities team must look to continuous dialogue with their customers to ensure the standards they have agreed are being met first time. This is why it is so important that the front-line staff have to take the responsibility to talk to their customers. Take a possible example: In a mail room the standard agreed with the customer is that all incoming post is delivered by 09.00 hours each morning. The customer is unhappy as the post seems to be delivered two hours late. The perception of the customer is that the service is bad due to poor delivery. However, the front-line person, the postroom operative, knows that the delay is in the customer's receiving area. If the postroom operative has the responsibility and takes it, he or she should start the process of communication with the customer and the facilities manager to redress the problem.

In this way the customer is made aware of a problem that cannot be resolved by the facilities team but can see that the standard is actually being achieved. This example again may sound almost too simple, but how many more situations are there where someone has not taken responsibility at the outset and the service to the customer or the customer's perception of the facilities team has suffered? Of course it may be that the delay is in the facilities area of responsibility, in which case the service should be quickly reinstated to conform to customer requirements.

The measure – the cost of quality

The cost of quality can be looked at in a number of ways. There is the short-term view that cutting down on the external painting of, say, softwood window frames is a saving, but the long-term loss is that the window frames need replacing sooner than if they had been maintained properly. Other simple examples could include having to retype a letter or repeat any job not done properly in the first place. Then there is dealing with customer complaints about shoddy workmanship or poor service. Some organizations have large customer complaints departments. This is throwing money at the symptom, not investment in eliminating the source of the problem. All these examples cost money, which is money unnecessarily spent. If the job had been carried out correctly first time there would not be any need for additional costs. Looking at ways of improving a task can lead to enormous cost savings. Sometimes it may be necessary to spend money up front with a view to recovering the investment later on through increased customer satisfaction. Individuals should be given more responsibility to look at their tasks and try to eliminate waste, therefore reducing costs even more. Given responsibility and delegated powers, individuals will often respond by taking the initiative on costs.

The standard – right first time

'To do it right, do it right first time' is an old saying which facilities managers should still try to achieve. The implications of not doing it right first time are to waste time for everyone (customers and the facilities staff); to put everyone under unnecessary pressure; to cost money; and to make customers dissatisfied with the service. No one likes doing a job badly – it demotivates them and their customers when not performed right first time.

The theme – continuous improvement

It has been mentioned before that TQM is a way of life. It is not a campaign with an end result to be achieved before moving on. TQM never ends. No matter how much we improve, competitors will continue to improve and customers will expect an even better service. Continuous improvement is about reviewing all services provided to customers on a continuous basis. Sometimes this can be daily, such as the need to receive post by a specific time. Other times it can be done when a contract is ready for renewal. It is also about asking the question: 'Am I taking responsibility for the quality of my work and the work I receive?' Continuous improvement is perhaps the most fundamental and difficult part of TQM to grasp. After the initial stock of TQM ideas, we need to ensure that things do not fall back to old ways. This needs good management support. Continuous improvement emphasizes the point that TQM is a way of life, not a campaign.

IMPLEMENTING TQM

Management support

It has already been mentioned that the facilities manager has to be seen to set the example that the team can follow. In many organizations TQM, in one guise or another, will already have started, so the facilities manager should have help and support in also implementing it. Sadly, however, a lot of organizations will not have started on the road to TQM and it will be up to the facilities manager to start the process.

Almost all of the basic principles of TQM are just good management technique. The problem is applying all of these techniques when inundated with work, when having to work to extremely tight budgets and when still needing to keep customers happy. But it is precisely these situations that make TQM an absolute necessity for today's facilities manager.

Quality circles

Quality circles are basically a team of people who work together and are empowered to improve the output of their team by identifying problems, investigating causes and providing solutions, the ultimate aim being to improve the services offered to customers. These issues can cover such diverse topics as the provision of footstools for switchboard operators to identifying a way to get rid of a power generator by utilizing other available resources.

It may seem that footstools etc. are not a quality issue, but it should be looked at this way: if it helps the front-line staff perform their job better at minimal expense it will eventually work its way through as contributing to a better service for the customer.

The savings made by the removal of the generator may be obvious. Not so obvious, but equally of benefit, is that it increases job satisfaction for the members of the quality circle and shows they are thinking about ways of improving the service and helping to reduce costs.

The facilities manager should be seen to support the quality circles. They should be encouraged to implement their own ideas, particularly the ones that cost nothing and improve the service. The manager should also ensure that any of the points raised by the quality circle which they cannot answer themselves should always receive an answer. It will be seen as one more example of the facilities manager supporting TQM.

Identify customers and clients

Facilities managers may sometimes get involved with conflicts outside their control, for instance where the client (the person who agrees the service and foots the bill) and the customer (the end user) seem to be at odds. An example of this could be air-conditioning plant in a building which is overdue for renewal and is giving poor and faulty service. The customers are very unhappy but the client baulks at the cost of replacing the equipment. Where does this leave the facilities manager? In a very difficult situation!

One solution is to make sure that the feedback to the client comes from the customer as well as from the facilities team. It is very important that the facilities team identify who is their client and who is their customer. Another example of the importance of this is where the teams are taking instructions from customers who haven't agreed the service. This is an area where conflict can easily arise unless client/customer identification has taken place.

Agreed service levels and standards

An outline of the service levels agreed with clients and the standards agreed with customers is as follows.

- Service levels are embodied in contracts between the facilities team and the client which are agreements to provide a certain service at a certain time in a certain way. The client is paying for these services (either directly through cost centre reporting or indirectly via overheads) therefore the agreement must be clear so that both parties know exactly where they stand.
- Standards are the detailed working agreements between the front-line facilities team and their customers (users). In many cases service level agreements and standards are agreed together, which helps understanding even further.

Service level and standards-setting can be illustrated with this example of a records management area in an insurance broker's office. Records retrieval is part of the office services function which in turn reports to the facilities manager. The overall service agreement between the client and the facilities manager is that a records retrieval system will be provided between 8.30 am and 6.00 pm Monday to Friday. The service standard, agreed at customer level, states that the files will be provided within one hour of request and wrongly 'pulled' files will exceed no more than half of one per cent of occasions during the month. The emergency service provides files within a quarter of an hour on request. Improved service level agreements and standards are achieved by delegating these as much as possible to the front-line team members. After all it is these individuals who are going to deliver the service to the client and the customer.

COMMUNICATION

Good communication is essential for quality improvement. Without appropriate communication virtually all the others aspects of TQM will fail. Good communication is about:

- keeping the facilities team aware of what is going on;
- keeping customers informed, especially if the agreed service levels are going to be difficult to meet;
- ensuring understanding;
- helping everyone work together.

SUPPLIER/CONTRACTOR INVOLVEMENT

How do you achieve participation in TQM by suppliers and contractors? In many organizations the majority of the service tasks such as cleaning, security and maintenance may be carried out by contract labour. The question is how do you get the contractors/suppliers to adopt a TQM approach which mirrors the in-house commitment? A good start is to ensure contractors/suppliers have either achieved the appropriate recognized ISO or BS Standard or are in the process of gaining them. These are fine as far as they go but they cannot guarantee you will receive a quality service.

What needs to happen is an education programme similar to the one which was started with the in-house facilities team. This ensures the contractors/suppliers know what service levels have been agreed, as the only way to ensure that contractors/suppliers deliver TQM in their work is to treat them as part of the 'internal' organization. Make them feel part of the facilities (which they are). The service levels and standards agreed between the contractor/supplier and the facilities team should always have a TQM basis. They should point out what is expected from the contractor/supplier and what is expected from the facilities team. This will go a long way to helping the contractors/suppliers buy into quality service. Of course, at contract renewal stage, when alternative contracts are being assessed, an integral part of the selection process should be to identify which companies already operate some form of TQM which follows a similar approach.

CONCLUSIONS

TQM is a way of life, not another campaign. The principles of TQM will help outline the background to providing a quality service. Quality circles are the key to success so the team needs to be involved a quickly as possible. Service level agreements need to be in place to ensure everyone knows what is required of them. There also needs to be good, positive communication through which everyone will have a greater understanding of how to deliver TQM. These are the keys to success.

SUMMARY

Adopting the quality managed facilities approach will ensure that facilities are tuned to business needs, that they consistently meet user requirements and are managed to provide a cost-effective service. Organizations with QMF will have an explicit quality policy and employ consistent processes to ensure achievement of well defined quality objectives. These policies and procedures need to be set out in a facilities plan. This sits alongside, but functions integrally with, the organization's business plan and shows how facilities can be developed to contribute to help achieve key business objectives.

Developing a quality culture for facilities management, and introducing the concepts and processes of quality managed facilities (QMF), requires a framework that can be described as a matrix for managing customers, service and assets at a strategic, tactical and operational level. There is no mystery about organizational performance improvement, though ingrained management practice and received 'wisdom' can create formidable barriers. However, an intensive project process, using simple and practical tools, can be shown to lead to dramatic progress in quality and cycle time and ultimately customer satisfaction

The techniques required to manage quality, value and risk, and the processes for developing and managing social and service partnerships, are key skills for facilities management and lead to conditions for continuous quality improvement. The application of total quality management can permeate every part of an organization and result in substantial benefits. To achieve this needs good and positive open communication whereby everybody feels involved and understands how to deliver.

BIBLIOGRAPHY

Essential reading

Crosby, P. (1984) *Quality Without Tears*, McGraw-Hill, New York.
Garvin, D. (1988) *Managing Quality: The Strategic and Competitive Edge*, Free Press, New York.
Munro-Faure, L. and Munro-Faure, M. (1992) *Implementing Total Quality Management*, Pitman, London.
Zeithaml, V.A., Parasuraman, A. and Berry, L.L. (1990) *Delivering Quality Service*, Free Press, New York.

Recommended reading

Camp, R.C. (1989) *Benchmarking*, ASQC Quality Press, Milwaukee, WI.
Deming W.E. (1988) *Out of Crisis: Quality, Productivity and Competitive Position*, Cambridge University Press, Cambridge, MA.
Ishikawa, K. (1986) *Guide to Quality Control*, Quality Resources Press, White Plains, NY.
Juran, J.M. (1992) *Juran on Quality by Design: The New Steps for Planning Quality into Goods and Services*, Free Press, New York.
Taylor, L.K. (1992) *Quality: Total Customer Service*, Century Business, London.
Willie, E. (1992) *Quality: Achieving Excellence*, Century Business, London.

Further reading

Alexander, K. (1992) 'Quality managed facilities', *Facilities*, Vol. 10, No. 2.
Alexander, K. (1993) 'Facilities management: the quality journey', in Alexander, K. (ed.), *Facilities Management 1993*, Hastings Hilton, London, pp. 80–8.
Beddek, P.J.H. and Kernoghan, D. (1992) *The Measurement of Quality in Buildings*, Centre for Building Performance Research, School of Architecture, Victoria, Wellington.
British Standards Institute (1992) *Quality in Action* BSI Quality Assurance, London.

Broh, R. (1982) *Managing Quality for High Profits*, McGraw-Hill, New York.

Cairns, G. (1994) 'Dynamic quality – the prerequisite of design innovation for industry', in Alexander, K. (ed.), *Facilities Management 1994*, Hastings Hilton, London, pp. 70–4.

International Facility Management Association (1992) *Quality Programmes in Facility Management*, IFMA Research Report No. 9, Houston, TX.

Scroxton, J. (1994) *The Future of Total Quality Management and ISO 9000 in the Hotel and Service Industries*, in BIFM Conference Proceedings, Saffron Walden, October.

Taylor, P. and Martin, R. (1994) 'Customer loyalty and clear policy top agenda', *The Financial Times*, 27 June.

Treleven, M. (1987) 'Single sourcing: a management tool for the quality supplier', *Journal of Purchasing and Materials Management*, Spring.

Xerox Corporation (1988) *Leadership through Quality*, Corporate Education and Training Handbook, Rochester, NY.

Value management | 6

6

OVERVIEW

Facilities economics is the study of man's attempts to create wealth through developing facilities. Bernard Williams (adapted)

As facilities managers demonstrate their particular contribution to the effectiveness of an organization by adding value to its operations, they use the core skills and tools of value management to contribute to that process. The value chain is the sequence of activities which build to generate the mix of products and services for the whole organization. The way in which value is added along the chain creates both the uniqueness of what is sold to the customer and also the costs. For a manufacturing organization for example, in its simplest form, it relates to the primary activities:

- designing the product;
- procuring materials;
- manufacture of the product;
- distribution;
- after-sales service;
- support.

The ability of an organization to create uniqueness and a particular cost structure – and therefore value – in its operations is influenced by the arrangement of its resources, including people, equipment, facilities, information systems, materials and the like. Considerable value can be added through tier effective organization and management.

Value management is an approach that promotes a systematic search for solutions that provide greater cost effectiveness, without compromising function or service. Pressures on organizations to control costs are ever present, especially where facilities are seen as an overhead on the business operation. For a facilities manager to achieve this control he or she must identify the significant, controllable and negotiable costs of operating facilities and delivering support services. Adopting value management, with its concepts and techniques of budgeting, operational control and auditing, will help to focus attention on providing facilities at the best cost to an organization rather than pursuing the least cost and ignoring other benefits. The process also contributes to teambuilding by providing participants with a better understanding of the perceptions of the other team members and thereby the organization as a whole.

When value management is applied to buildings and property, the concepts and techniques of identifying and managing costs, defining valuation parameters, capital charging and controlling overheads (including their impact on managing the physical environment through 'premises audits' and 'asset management') all contribute.

Added value is a commitment to partnership. As such it needs to be a partnership at top board level between the core interests of the business and the service interests that support them, as represented by the value management champion for the organization – the facilities manager. Top companies recognize the value that can be added by effective management and exceptional service, so they organize facilities management to enable fulfilment at a business rather than a technical level.

Facilities value management looks at the ways in which value is added in an organization through facilities management and suggests that if facilities are tuned in they help meet business objectives. The conclusion from this is that facilities can be effectively managed only if seen as an integral part of the enterprise.

Keynote paper

Value management
Oliver Jones

INTRODUCTION

In order to appreciate the nature of value management it is necessary to have an understanding of just how the facilities manager fits into the overall organization and its corporate strategies. Put briefly the FM input should begin along with the inception of the organization. Where this coincides with the procurement of a new building, the input will arise at the outset of the design stage. With existing property it will be at the operational start-up phase. Thereafter, during ongoing occupation, and particularly during periods of organizational change, input will be continuous and sometimes intense.

MAINTAINING CONTEXT

When first examining the issue of value management it is critical to look at the context rather than the detail. The context must be maintained, as the analysis and understanding of strategic decision-making is a key element of the development of an appropriate and relevant facilities strategy. If this is not done, justification of any budget, or control of the costs within that budget, can become difficult if not – and this is more important – irrelevant. Organizations, and facilities managers within them, generally aim to focus on achieving 'cost-effective' facilities management by managing the optimum mix of in-house and contracted support. However, it can sometimes be debatable, not least by those providing the service, just what 'cost-effective' means in this context.

There are three key areas concerning value to consider against this backdrop:

- strategies for adding value;
- techniques for managing value;
  mechanics for demonstrating value.

The facilities manager's task, in its broadest context, has to be to service and support the organization at the most effective cost, having regard to other non-financial objectives of the organization. The importance of strategy is therefore evident. It is the responsibility of the facilities manager to develop a clear strategy for resourcing the service that is based upon organizational objectives. This has to be done before entering into the detailed realms of absolute cost reduction and any related matters.

ORGANIZATIONAL OBJECTIVES

Organizational objectives for the facilities function at all levels can include a combination of the following:

- to gain maximum value for money for any work on the maintenance or rehabilitation of the buildings;
- to achieve time savings when procuring this work in order to allow

the buildings to continue to function efficiently and maintain maximum benefit for the users;

- to ensure high quality workmanship is achieved at all times;
- to determine how the image may be used to increase the profile of the organization;
- to ascertain how financial benefits – in the case of property particularly – may be gained through possible restructuring.

The process of determining an effective facilities strategy requires an understanding of the present situation (covering property, maintenance and other support services such as cleaning, security and catering) and the present and future objectives of the organization (with particular regard to the impact of these on the estate and support services). The strategy must address not only the structure of the organization and its personnel, but the value of the work, the changing expenditure profiles, the procurement route options, etc.

The key requirement is not only to identify how to develop a value-adding facilities strategy, but to do so in the context of recognizing that factors other than cost alone are of significance. Achieving the lowest cost is of little relevance if the principal goals are not accomplished. Getting value out – the strategies of performance management, applying incentives and quality management through partnership sourcing – is the focus of activity once an effective facilities strategy has been determined.

THE COST FUNCTION

In looking at the linkage of cost and strategy development, the following two questions are often asked:

- Is cost saving a reason for outsourcing?
- Under what circumstances would organizations expect to achieve savings, and where are these generally found?

As the subject of outsourcing is very wide-ranging and specific to individual cases, it is useful to address these questions from the perspective of the decision-making process. Facilities management practitioners can then gain greater insight into the principal options and issues involved as follows.

BACKGROUND TO OUTSOURCING

In order to make sensible decisions about outsourcing it is useful to look at some of the reasons that lead to its use.

- Competitive pressures force attention on overheads or, in the public sector, erratic public expenditure profiles from year to year.

- Skill shortages – specific in the case of many professional support staff and general in the case of management staff – force organizations to look towards contracted support on the grounds that in-house managers should focus on the core business.
- Manpower shortages in general are a major factor. As a consequence of the decline in the availability and number of quality graduates, organizations are forced to look at other sources for their non-core businesses.
- Flexibility is important, particularly in the case of unpredictable markets which require both base costs and fixed overheads to be low. External sources can provide this flexibility, although in many instances it could be achieved through creative employment contracts.
- Specialist knowledge is required for modern technologically advanced buildings and, in many instances, this has outstripped the ability of a single organization to adequately service its own facilities.
- Trends are followed, e.g. pursuing the contracting-out route because competitors are or, in the case of the public sector, pursuing a route due to government and related pressures to market test.
- Lower cost or affordability is possible, i.e. obtaining suitable resources from the market at lower cost.

The cost argument is the one that is most often used. However, it is frequently the case that direct comparisons are not made. The principles on costs are well summarized by the graph of resource/asset specificity given in Figure 6.1.

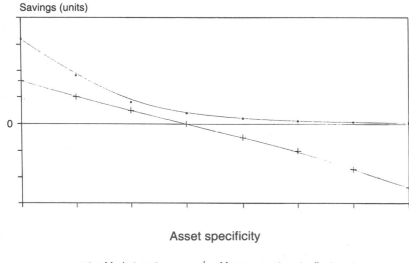

Figure 6.1 Asset specificity

The graph effectively illustrates a commonsense principle – that where any resource is highly specific to the organization, i.e. there is a limited external market in the provision of the service, and the organization can accommodate the cost through appropriate salary structures, in-house provision is likely to be the preferred route. In other circumstances, contracted provision is likely to be favoured.

While external market costs can tend to be lower for a supplier that specializes in a given service, the internal management and procurement costs of the buying organization still apply. For non-core services such as cleaning, catering and security, the contracting route could be the financially preferable option, provided an appropriate management structure exists internally, the costs of which still have to be considered.

The reasons given by organizations for not using contract support include the following.

- Concerns are raised by users (customers) over the ability of suppliers (providers) in respect of quality, availability, capacity commitment, continuity, mistrust, etc. These concerns are often treated as though they are unchangeable, but if known about and understood they can be specifically addressed when preparing tender documentation.
- Risks, in the specific sense of security or of internal industry relations, etc., are concerns frequently raised, often as gut reactions. The facilities manager's task is to see beyond this and identify exactly what the organization requires overall. He or she needs to know how to effectively tap the market in the service of the organization with minimum risk, both specifically and generally.
- The cost of effective procurement offset against the benefits of market competition is a concern sometimes quoted. The costs of effectively accessing the market need to be assessed.

Although there are always good reasons both for and against outsourcing, the aim should be to remove the specific prejudices held by individuals. The focus must be on the organization's core objectives, rather than any facilities results. If the aim is to ensure that the organizational objectives are met through the use of an appropriate risk analysis technique, the most suitable contracting strategy for the organization can be identified.

CONTRACTING STRATEGY DEVELOPMENT

One approach to developing a contract strategy is to take each support service and test whether the principal objectives of the organization are influenced by that service. Where the failure of any of these support services puts an objective at risk, the potential contracting mechanism must be redesigned to counter such a threat. In the event that a suitable contracting route cannot be structured, then the in-house alternative will need to be developed. Operational effectiveness and cost-effectiveness will then become common goals.

As previously outlined, relative risk is a significant issue. Relative risk has to recognize that the damage resulting from the poor selection of a low-cost bidder is likely to totally outweigh the difference in cost to the next lowest bidder or the alternative of in-house provision. It follows that total expenditure must be a principal feature in the selection process.

Why are strategy and risk relevant to resourcing a facilities management service? The answer is that without a clear strategy, and a full assessment of risk, no resultant facilities operation would adequately service the needs of the organization. Also, the tendency to dwell on the detail without addressing the fundamental overall structure can often result in lower quality and higher cost than is optimal for any given situation.

The development of a strategy must, as in all cases, address three issues; firstly how the current facilities are provided; secondly what the optimum facilities structure will be; and thirdly how to move between these two points. This may sound simplistic, but it is often this fundamental element of an outsourcing strategy that is ignored.

COST CONTROL

Cost control is necessary at both the contract level and the asset level if value is to be recognized. The control at contract level is important for operational reporting and management. At asset level it is essential for later analysis and trend evaluation. Effective procurement is also a fundamental element for success in value management. Risk has to be set off against cost certainty through an appropriate contracting mechanism and, at the same time, the aim must be for goal congruency between the participating contractor and the employing organization. The facilities manager should seek to establish what tolerance exists within his procurement strategy. Lump sums can give cost certainty, but at what cost? Quality? Response times? Effective procurement requires effective management thereafter and appropriate procedures having regard to the contract risk should be adopted. Value-adding procurement has to be seen in this light and must cover the following.

- **The task specification**. Clearly identifying the responsibilities and expertise required and giving all the relevant information to enable the contractor to understand both the scope and the methodology required. Decisions have to be made between prescription specifications and performance specifications, recognizing the advantages and disadvantages of each.

- **The contract strategy**. Whether a time-based, performance-based or activity-based approach should be adopted has to be determined, and whether specific performance measures should be incorporated and how. The approach must seek to protect the interests of the employer adequately without imposing undue or unacceptable risk on the contractor.
- **Marketing intelligence**. Being in touch with the market, understanding the industry segmentation and recognizing the skills of those within the various segments is the key to introducing competition and bid efficiency.
- **Programming**. Providing adequate time for participants to analyse and cost documentation, and subsequently to mobilize and commence operation, is an often ignored opportunity for risk removal.

To demonstrate cost-effectiveness within this environment requires the facilities manager to have the ability to systematically collect and record expenditure in an accurate and readily accessible manner. It is not possible to control elements that cannot be measured in some way, even those with a subjective aspect to them. If information on how much has been spent on a particular element is not available, then it is not possible to assess the effect of any efforts to reduce or control expenditure.

Dependency on central accounting functions can in many instances introduce delay and aggregation of cost. The form in which the facilities manager requires information must have regard for the need to analsye information in the necessary form, whether it be, at a minimum, at billing level, element level or contract level. In this way, the facilities manager may be able to demonstrate to the same effect as the core business lines the past record and benefits of control. The emphasis throughout is on the importance of having a strategic review, and within that a tactical implementation plan. In evaluating the success of such strategies, the importance of data and its accessibility is critical.

A very low cost way of managing a property can be, for example, using *Yellow Pages* and a crisis telephone call. However, expending budgets at low cost in a responsible way requires cost-effective management. This can come from maximizing the support services obtained for a given cost – the budget and applying them in a way that has regard to the organization's requirements.

CONCLUSIONS

There is no single right answer for a value-based approach to estate operations and facilities management. Empirical data is essential to demonstrate achievement, as is maintaining an adaptable outlook to the requirements of different types of estates. Where a single right answer is claimed, it usually demonstrates a very narrow view of the impact that the facilities organization has on the organization that it is serving.

Value comes through professional management and the procurement of support services at the outset, and through demonstrating the impact of risk in close accord with the strategic direction of the organization. It also comes when addressing the ultimate trade-off that exists between cost and quality when looking at the tendering process.

The precise tailoring of a contracting structure to a given organization is where the value-adding service of the facilities manager is best applied. 'Off-the-shelf' solutions will rarely be ideal. If facilities management is to continue to go from strength to strength, it must be understood and recognized that the facilities function is a service to the core business and therefore an integral part of it, and not, as sometimes seems to be the case, an entity that exists for the benefit of those participating in it.

SUMMARY

Value as a concept is rather meaningless without a full understanding of the context in which it occurs. By necessity, the process of value management cuts across boundaries – so facilities can only be managed effectively if they are seen as an integral part of the enterprise. To monitor their value and utilization requires integrated measures of the 'throughput' over time against the costs of facilities.

The importance of a clear strategy and of a contracting structure tailored to a given organization is paramount, in that they highlight the trade-offs between cost and quality in the tendering process. Value comes through professional procurement and management and through the provision of data to demonstrate achievement.

Information is the key to the value question. Far-reaching facilities management decisions can only be instituted because of the existence of an accounting structure that enables the FM to demonstrate the cost advantages of any proposals, reforms or otherwise. Value management is not helped if business accounting systems do not disaggregate financial information in a form which supports the facilities manager. For this to happen, facilities performance indicators – or benchmarks – need to exist such that performance can be compared across organizations and their buildings. Concern about the collection and exchange of performance information between organizations in order to establish independent, credible and relevant benchmarks is important as it helps get value from the other sectors. Value comes through professional management and procurement, as well as demonstrable risk avoidance, in close alliance with the objectives of the organization.

BIBLIOGRAPHY

Essential reading

Band, W.A. (1991) *Creating Value for Customers*, Wiley, New York.
Bone, C. (1993) *Value Analysis in the Public Sector*, Longman, London.
Clift, M.R. and Butler, A. (1995) *The Performance and Costs-in-use of Buildings: A New Approach*, BRE Report, Building Research Establishment, Garston.
Williams, B. (1994) *Facilities Economics*, Building Economics Bureau, Bromley.

Recommended reading

Bone, C. (1992) *Achieving Value for Money in Local Government*, Longman, London.
Flanagan, R., Norman, G., Meadows, J. and Robinson, G. (1989) *Life Cycle Costing: Theory and Practice*, BSP Professional Books, London.
Fowler, T.C. (1990) *Value Analysis in Design*, Van Nostrand Reinhold, New York.
Kelly, J. and Male, S. (1993) *Value Management in Design and Construction: The Economic Management of Projects*, E & FN Spon, London.
Layard, R. (1972) *Cost Benefit Analysis*, Penguin, Harmondsworth.
Stone, P.A. (1980) *Building Design Evaluation: Costs-in-use*, E & FN Spon, London.

Further reading

Apgar, M. (1993) 'Uncovering your hidden occupancy costs', *Harvard Business Review*, May–June.
Ashworth, A. (1988) *Cost Studies of Buildings*, Longman Scientific & Technical, London.
Brandon, P.S. and Powell, J.A. (1984) *Quality and Profit in Building Design*, E & FN Spon, London.
Centre for Facilities Management and SJT Associates (1994) *Facilities Costs and Trends, FACTS '94*, CFM, University of Strathclyde.
Cooper, R. and Kaplan, R. (1991) 'Profit priorities from activity-based costing', *Harvard Business Review*, May–June.

Dell'isola, A. and Kirk, S.J. (1981) *Life Cycle Costing for Design Professionals*, McGraw-Hill, New York.

Editorial (1990) *Facilities Management Economics*, Facilities Management International Conference Proceedings, Centre for Facilities Management, University of Strathclyde.

Fallon, C. (1971) *Value Analysis to Improve Productivity*, Wiley, New York.

Gage, W.L. (1967) *Value Analysis*, McGraw-Hill, New York.

Zimmerman, L.W. and Hart, G.D. (1982) *Value Engineering: A Practical Approach for Owners, Designers and Constructors*, Van Nostrand Reinhold, New York.

7	# Risk management

OVERVIEW

It is a world of change in which we live, and a world of uncertainty. We live only by knowing something about the future, while the problems of life, or of conduct at least, arise from the fact that we know so little. This is as true of business as of other spheres of activity. The essence of the situation is action according to opinion, of greater or lesser foundation or value, neither ignorance nor complete and perfect information, but partial knowledge. Prof. Frank Knight

Like quality and value management, risk management is the responsibility of everyone in a business enterprise. In most organizations the processes of risk management have assumed strategic importance. Risk management means a course of action planned to reduce the risk of an event occurring and/or to minimize or contain the consequential effects should that event occur. The help which an organization arranges to manage any loss-producing event which occurs – pre-emergency, emergency handling and recovery – is contingency planning, whereas the process of restoring operations and minimizing the loss associated with an occurrence is disaster recovery.

Every business decision involves a range of risks – to the workers, to the 'workflow' (i.e. the process), to the environment, to property and, ultimately, to the organization's financial performance. Particular attention has to be paid to identifying the extent and nature of these risks, to their elimination or control, and to managing their impact on the business. This involves allocating responsibilities and planning for contingencies and disasters.

The key skills required for effective risk management are risk awareness and effective communication. If to these is added the ability to prioritize risk control measures and to persuade the finance director of the need to spend on these non-profit-making areas, often in a time of budgetary constraint, this gives an effective manager of risk. A facilities manager can then carry responsibilities for risk and provide a plan of action for facilities risk management.

In all organizations the extent of the business risk borne by the facilities team is very significant. Increasing legislation and litigation has raised the awareness of senior management to the need for effective control. Accountability for health and safety, for environmental impact and for financial viability lies with senior managers, and they will seek to delegate responsibility for the effective management of risks to those directly responsible for providing the service.

If risk management is the responsibility of everyone in the business enterprise, it is necessary to organize and manage to encourage everyone to play their part. Training at all levels is essential. People must be aware of the risks and of the steps which can be taken to prevent threats becoming a reality. Ultimately, the viability of the organization and the security of each individual's job may be 'on the line'. In facing the basic concepts of risk management and their application to the field of facilities management, the facilities manager has to be able to identify the main classes of relevant risk and to apply models for determining the exposure of an organization to risk and for identifying the level of control that may be exercised over particular risks.

Keynote paper

Practical risk management
Roy Dyton

INTRODUCTION

Risk management is a tool that can be applied to an organization as a whole or to specific areas of risk within it, and it can be adapted to meet particular circumstances. Some organizations need complex statistical analysis as a means of estimating levels of risk; some concentrate on physical inspections together with human motivational techniques to encourage attitude changes, while others concentrate on risk financing techniques to minimize their dependence on the conventional insurance market. Each has its place. Perceptions of risk vary, with some people, for example, enjoying the thrill of hazardous pursuits, while others are more cautious. Corporate attitudes to risk, national and international, will also vary as they are inevitably moulded by individuals: some will be risk takers, others risk averse – a retailer insuring against burglary but not pilferage, for example. Attitudes such as this are common and often reflect both historical as well as individual values.

DEFINING RISK

Clearly risk can have different meanings but the normal understanding is that the risk could actually happen, and its consequences might not be pleasant. Definitions of risk must always relate to the risk of something happening in a specific time period. A variety of terms are in common usage which are often meaningful to those active in the business within which they are employed, but they can lack precision when it becomes necessary to examine them critically. For example, risk can be used to mean:

- a hazard or unsafe practice;
- a peril capable of being insured, i.e. fire or storm;
- the subject matter of an insurance policy;
- a statistical probability;
- loss potential as assessed by an insurance surveyor – estimated maximum loss;
- the actual value at risk.

However good past loss data is, it cannot be assumed an accurate predictor of future events. The world we live in changes and it is easy to overlook the influence of the effect on attitudes and values. Risk must always be considered within the constraints and controls of the social setting within which it takes place, as those involved with the East/West German unification, the *Herald of Free Enterprise* disaster and so on can testify. The challenge for the business manager is to determine what level of risk is 'acceptable'. Hazard reduction involves expenditure. Will the money so spent yield a positive or a negative return? How can such a sum be calculated?

QUANTIFYING RISK

Risk can be considered both from the view of the social scientist as well as from a totally scientific viewpoint. The former takes account of the attitudes and the culture within which people work, whereas the latter assumes that all the factors are capable of precise calculation. Evidence from actual case studies shows that this is not always so.

Risks may be categorized as follows:

- those for which identified statistics are available;
- those for which there may be some evidence but where the connection between the suspected cause and consequence cannot be established;
- expert assessment of probabilities for events that have not yet happened;
- non-expert assessment of probabilities of events that might happen; and finally
- the only technique available in many business situations: making a probability assessment which depends entirely upon subjective judgement or gut reaction.

Resolution of such situations can be achieved by the traditional decision-making process, which entails examination of the amount exposed, the availability and cost of insurance, the possibility (and expense of) risk elimination or containment, the significance of action to corporate survival and the liquidity position of the organization itself.

THE CORPORATE ENVIRONMENT

What choices are open to an entrepreneur operating within a corporate environment? An organization that aims for a given expansion from current worth in a given time can only predict this with complete certainty in a totally risk-free economy. In such a case insurance etc. is unnecessary. On the other hand expansion would be less than anticipated because, although saving on insurance, all risk event costs would need to be financed from internal resources. Alternatively the organization might have had better growth long term if it had effected insurance, thus transferring risk to a specialist insurance body, or if it had invested in loss control technology. A combination of these two methods would give best performance but would still be below optimal position as both involve an outflow of corporate assets so depleting corporate net worth. Such precision is not practical, of course, but it does highlight the economic challenge which all businesses face, namely how to maximize future net worth. It also shows that cost sometimes needs to be measured against the alternative which has been foregone (opportunity cost), and not just in monetary terms.

Risk cannot be divorced from the financial position of an organization. The cost of an event will be more serious for one with an overdraft than for one with credit. An organization will have a number of financial goals and an unexpected accident may adversely affect all or some of them. The timing of money movement after a disaster is important. The organization is no worse off in financial terms immediately – not until the loss creates a knock-on effect by way of reduced income because of lower sales, a replacement building or loss of customer or investor confidence. Consequently, the facilities manager with risk responsibilities needs a range of skills to appreciate the role of finance in the successful operation of an organization. Business functions around a budget and is the usual tool by which financial performance is judged internally. In a large or small organization the budgetary mechanism is needed to monitor corporate performance. Any unexpected factor, or risk, which throws the agreed budget off course is an indication that the corporate plan has gone awry.

THE RISK MANAGEMENT SYSTEM

Risk management is concerned with the management of fortuitous events which would have an adverse impact on the balance sheet and cause asset depletion and/or income reduction. It is defined here to mean the identification, measurement and economic control of risks that threaten the assets and earnings of a business. No reference is made to liabilities since they necessarily reduce assets. A risk management system includes the following elements:

- **hazard** – a situation that in particular circumstances could lead to an adverse event;
- **risk** – the probability that a particular adverse event occurs during a stated period of time;
- **event** – an adverse event which produces harm;
- **damage** – constitutes harm or loss to people and includes loss of quality inherent in any physical entity.

Such a system is illustrated diagrammatically in Figure 7.1.

Risk management is concerned with a variety of practical issues such as:

- determination of values at risk;
- compliance with legal obligations;
- identifying threatening contingencies;
- accident prevention;
- occupational health of employees;
- communication and behaviour influence; and
- alternative ways of risk financing.

The application of risk management will vary between organizations, so, while emphasis and understanding will differ, the whole will still be wider than its parts. For example, with insurers, risk management is a

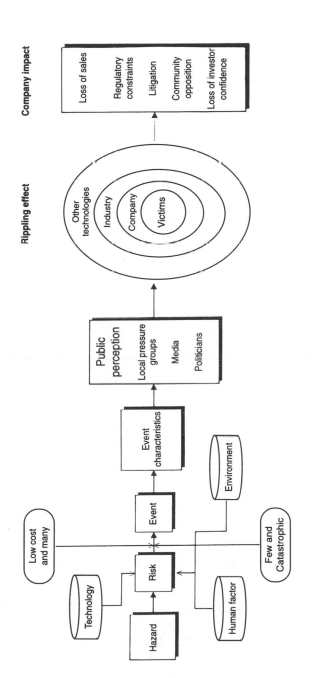

Figure 7.1 The risk management system

means to contain claims and improve profit, whereas with consultants the emphasis could be on health and safety, seen then as risk management, not a component of it.

Risk management covers the totality of all events which can have a downside effect on the balance sheet and it entails examination of:

- their potential causes and consequences;
- their magnitude and frequency;
- any historic loss pattern;
- hazard cause and eradication;
- loss containment measures;
- risk retention or transfer methods;
- risk financing methods, and
- the organizational patterns underpinning any risk management strategy.

Essentially, concern here is with downside or 'pure' risk – events which have only negative effect. With 'commercial' risks events can give both positive (gain) and negative (loss) results.

Risk management should always be commercial in its goals and application. Some basis for justifying capital expenditure to eliminate or reduce risk is essential. Clearly, it is unwise to wait for an event to do this. Equally, no organization can afford to spend indiscriminately on risk improvement without tangible evidence to demonstrate the benefits. Ultimately, it is necessary to decide what is an acceptable level of risk which an organization (or society at large) should bear. The presumption is, then, that the cost of doing nothing (i.e. accepting the risk) is less than the expenditure that would be entailed in undertaking some form of preventative or containment measures. The risk management process is illustrated in Figure 7.2.

Risks may be categorized as:

- those for which identified statistics are available;
- those for which there may be some evidence, but with no established connection between the cause and effect; and
- expert assessment of probabilities of events that have not yet happened.

For example, there are government statistics which enable reliable comparisons to be made which permit predictions of reasonable accuracy. This is quite different from the day-to-day problems facing many organizations which need to decide what to do about a risk when such a possibility occurs infrequently. Making decisions under conditions of total uncertainty is one of the fascinating challenges of risk management.

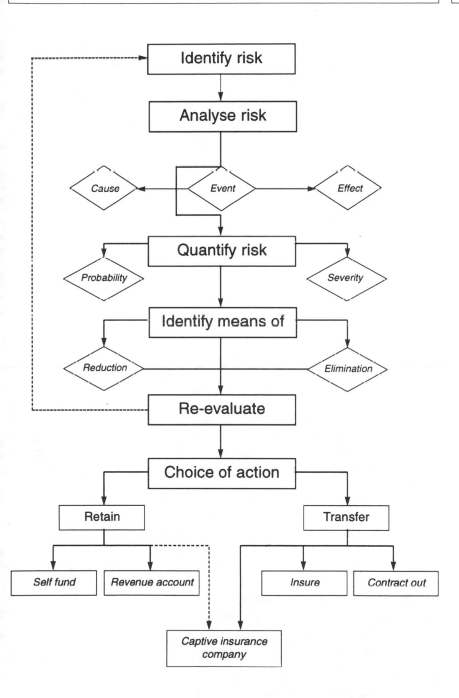

Figure 7.2 The risk management process

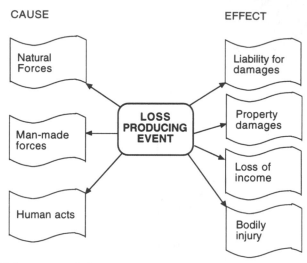

Figure 7.3 Risk event analysis

RISK ANALYSIS

Risk analysis, a component of the risk management process, deals with the causes and effects of events which cause harm, as illustrated in Figure 7.3. The goal in risk management is precise and objective calculation, which is the expression of the solution to a statistically based problem. If this were to be achieved then the decision-making process would be more certain. However, there can be significant personal influences which may affect the accuracy and reliability of scientifically based data. This is partly because an analysis depends on some subjective judgement, and partly because it may be necessary to build certain assumptions into the analytical model simply because there is no data available.

The likelihood of an adverse event happening can be estimated both with regard to its frequency and severity. Such an event can take on a wide variety of characteristics which will have varying degrees of seriousness, depending upon the nature and extent of the damage and the perception which others may have of it. Sometimes public perception of an adverse event may influence the feeling of commercial discomfort felt by management. All of this can trigger a rippling effect and once risk analysis advances to this extent, it is increasingly dependent upon judgemental assessments owing to the absence of statistical data. This is not to deny the value and necessity for such an approach. The potential pressure that can be exerted on a company's business plan by adverse customer response is a force that cannot lightly be ignored in today's economic climate.

Organizations will have different motives for undertaking risk analysis. These will include:

- to know what risks exist;
- to be a good employer;
- to comply with legal obligations;
- to do something about identified risks;
- to save money;
- to ensure survival.

People have different perceptions of risk too. When querying different hazards the answers might differ according to who was asked. However, many events are designed in the knowledge of the hazards involved. This is risk management in practice but it works so well we often overlook it for what it is. Perception therefore needs to be used with care.

RISK ANALYSIS TECHNIQUES

Various techniques together with their application in a practical working environment are discussed below as follows: historical data, question-naires, checklists, physical inspection, HAZOPS, fault-tree analysis and failure mode and effect analysis.

Historical data

Historical loss data is not always available or appropriate in practice:

- waiting for evidence can be wasteful;
- average losses can fluctuate unless the available data is extensive;
- data may be influenced by statistically predictable socioeconomic factors;
- rigid adherence to accounting-based criteria can overlook hidden costs.

All these factors can be present and investment decisions need to take account of both the direct historical costs and hidden costs.

Questionnaires

What type of information should we be seeking? The form might include:

- assessing the maximum exposure for planning or insurance purposes;
- assessing expected loss frequency;
- measuring risks against independent quality standards.

Questionnaires can be useful as a means of structuring thoughts, as well as a means of assisting others to provide meaningful information with a minimum of effort. More detailed ones, appropriate to the specific needs of individual companies, are often necessary. Questionnaires aim to provide stimulus to the critical appraisal of all aspects of the problem under investigation. Their successful use depends upon the expertise and judgement of the investigator.

Checklists

Checklists can be useful as a means of guiding the facilities managers on hazards to be regularly checked and are useful as evidence provided they are a faithful record. However, it is often necessary to consider other techniques where management commitment can be won and maintained.

Physical inspection

Physical inspection is an essential part of any risk analysis exercise. Experienced human eyes, ears and instinct produce quality information. Personal contact will enable the facilities manager, in his or her risk management role, to:

- see it as it really is, 'warts and all';
- make a comparison with other locations and with his or her own subjective expectations;
- influence managers and staff by the interest shown and said.

So it is important to judge whether the value of the higher quality information forthcoming will outweigh its acquisition costs. A survey of the premises will seek to obtain the type of information detailed above for questionnaires but in greater detail and with more precision. Inspections are jobs for specialists.

HAZOPS

'Hazard and operability studies' are a qualitative approach to extremely complex risk identification, using manageable-sized components applied at the planning stages of a project. Organizations are broken down into parts so that associated hazards can be identified, and the study is conducted within a structured framework concerned with four questions:

- the intention of the part under investigation;
- deviations which may happen from the declared intention;
- the causes of such deviations;
- their consequences.

It is essential to provide adequate definition of all processes/parts incapable of providing commercial guidance.

Fault-tree analysis

This method enables risk to be assessed in a quantitative manner, even though some of the probability assessments may need to be subjectively based. The technique requires more detailed analysis of processes rather than the broad effect it has on risk. In order to reduce the risk of an adverse event happening, fault-trees allow the statistical advantages of substituting alternative equipment or procedures to be calculated.

Failure mode and effect analysis

This methodology, which prioritizes improvements by ranking, involves a carefully defined examination and detailed analysis of each process to identify cause, effect, severity, frequency, etc. By a simple scoring system it is possible to determine which requires re-examination in order to reduce its risk score.

THREAT/EVENT ANALYSIS

All businesses possess but two essential attributes: assets and the income that is generated by their efficient deployment. Risk management is concerned with their economic safety in terms of assets, income and threats. Profiling can create a 'pen-picture' of the company which involves identifying hazards. Company reports and accounts, publicity and sales brochures, organization charts, process flow-charts, insurance registers, property registers and so on all contain information as possible profile sources. Once compiled the challenge is to test the initial assumptions to confirm the nature and extent of the threats which exist. Ideally, information should be obtained by face-to-face discussion with the relevant senior managers coupled with a physical inspection of the major sites.

Initially, it is important to focus attention on key issues – those risks that could threaten the survival of the company through costly repetitive losses. Analysing events is easier than attempting to analyse all the potential causes of risk. No specialized training is required and attention is focused on the multiple consequences of a single event. 'What if' questions are reasonably easy to frame and remedial action to contain the effect is often easier than eliminating the cause. However, this is not a technique capable of being determined in advance; rather it will need to be undertaken as more information becomes available as enquiries progress.

Event analysis looks at the exposure side of the risk equation in order to assess just how much finance could be at risk in a given situation. It is particularly relevant to the facilities manager who is responsible for the provision of specialized facilities to a new customer. By clearly establishing agreed service standards it is comparatively easy to identify possible downsides and then plan measures designed to ensure they do not exceed tolerable limits. Necessary measures might include:

- operational manuals;
- improved operator training;
- warning devices;
- back-up equipment for critical components and systems.

RISK TRANSFER

The insurance option

Insurance is the principal, well established, risk transfer mechanism. Its advantage is that an unknown risk can be exchanged for an identifiable premium – a financial transfer in a reputable industry. However, insurance is not the panacea it once was. Premiums can fluctuate and are expensive. Capacity is diminishing and insurers are becoming more selective. It is not a 'sleep-easy' nor a charitable option and insurers, too, need to make a profit if they are to survive! Insurance is a contract of the 'utmost good faith' and the normal rules of 'let the seller beware' do not apply, but it is exceptionally easy for policies to be inadvertently invalidated – so beware! Poor trading results in equivocation on claims for commercial considerations. Careful investigations and planning of an insurance programme for an organization requires skill, so watch that items are adequately covered. Policies invariably contain conditions and warranties which require absolute compliance by the policyholder – make sure these are understood and communicated to all relevant personnel. Organizations are prone to avoid thinking the unthinkable and are caught out with inadequate cover by a catastrophe loss. Insurers become insolvent just like any other company, so again take care!

The legal process

Organizations, given lawful intent, can decide how they wish potential liabilities for unsatisfactory contractual performance to be apportioned. Resorting to the law, of course, is not always beneficial and the pitfalls include the following:

- poor opponents may have insufficient money to give satisfaction;
- drafting accuracy requires experts to avoid unexpected results;
- as the law evolves so long-established decisions can be overturned;

- unfairness may result in abuse – now reduced by the Unfair Contract Terms Act 1977.

Businesses are allowed the freedom to regulate their transactions as they prefer, so long as they accept liability for bodily injury and are able to satisfy the test of reasonableness. There are a number of potential pitfalls in the way of those who wish to rely upon the use of contract terms (often referred to as 'hold harmless' agreements) as a means of risk transfer. Some companies have used them to good effect in order to protect their commercial position, with standard contracts drafted so as to exclude or limit the consequences of any contractual breach. One practical solution to difficult risk transfer problems (where insurance is expensive or unavailable and the potential financial risk unacceptable) is to discontinue the task and engage an independent contractor to perform it on your behalf. This is outsourcing and it can usually be a perfectly acceptable solution to a difficult problem.

CONCLUSIONS

In practical risk management, historical data can well be an accurate and sufficient predictor of future events. Nevertheless, we live in an age when change is the norm, so it is unwise to assume that the hard evidence of the recent past cannot be modified by incipient trends scarcely yet discernible. Change can undermine well established patterns of risk, necessitating periodic and stringent reappraisal of the assumptions underpinning the corporate business plan. Technological development, for example, frequently necessitates substantial capital investment, but development times and complexities can have devastating consequences for future profitability if the original budget has been poorly prepared, or if project overrule enables a competitor to secure a larger market share. Risk can be influenced by social, economic, technological, management and legal factors, sometimes to a quite wide-ranging extent.

SUMMARY

All management decisions involve future prediction and therefore uncertainty. In the face of uncertainty there are two factors – knowledge and response. The knowledge is what a person knows or should know, and that includes both the law and the factual situation. The response which follows naturally from this is to be responsible for what is needed - to speak, to act or to wait. This applies to all risk handling, from the immediate response necessary in a moment of crisis, to consideration of strategic planning for risk management throughout a whole organization.

Concepts of risk management can be related to a legal framework for facilities management. The creation and declaration of the law, and then judgement and enforcement of the right response, apply these concepts of knowledge and responsibility. The exposure of the public sector to risk has been dramatically increased through the removal of Crown immunity and the introduction of Crown indemnity, together with limitations on the purchase of insurance. Quality management and environmental management systems also include procedures for effective identification, assessment, control and management of risk.

Risk management is an integral part of the long-term approach of facilities management and aids organizational survival. Ignoring or underestimating risk can lead to catastrophe, so the nature and processes of risk management must be understood. They are the foundation for the application of assessment techniques to threats to the organization as a whole, and to component processes within the organization.

BIBLIOGRAPHY

Essential reading

Bannister, J.E. and Bawcutt, P.A. (1981) *Practical Risk Management*, Witherby, London.

Dickson, G. (1989) *Risk Analysis*, Witherby, London.

Knight, F.H. (1921) *Risk, Uncertainty and Profit*, Houghton-Mifflin, Boston, MA.

Thomas, H. (1990) *Risk, Strategy and Management*, JAI Press, Greenwich, CT.

Recomended reading

Boyadjian, H.J. (1987) *Risks: Reading Corporate Signals*, Wiley, New York.

Dowie, J. and Lefrere, P. (eds) (1980) *Risk and Chance*, Open University Press, Milton Keynes.

Flanagan, R. and Norman, G. (1995) *Risk Management and Construction*, Blackwell, Oxford.

Health & Safety Executive (1989) *Quantified Risk Assessment: Its Input to Decision Making*, HMSO, London.

Moore, P. (1983) *The Business of Risk*, Cambridge University Press, Cambridge.
Singleton, W.T. (1987) *Risk and Decisions*, Wiley, New York.

Further reading

Alexander, K. (1992) 'Facilities risk management', *Facilities*, Vol. 10, No. 4, April.
Finch, E. (1992) 'Risk and the facilities manager', *Facilities*, Vol. 10, No. 4, April.
Fischhoff, B. (1983) *Acceptable Risk*, Cambridge University Press, Cambridge, MA.
Health and Safety Executive (1991) *Successful Health and Safety Management*, HMSO, London.
Hertz, D.B. and Thomas, H. (1983) *Risk Analysis and Its Applications*, John Wiley & Sons, New York.
Hertz, D.B. and Thomas, H. (1984) *Practical Risk Analysis: An Approach through Case Histories*, John Wiley & Sons, New York.
Hollman, K.K. and Forrest, J.E. (1991) 'Risk management in a service business', *International Journal of Service Industry Management*, Vol. 2, No. 2.
Keeling, R.J. (1992) 'Legal risks in sick buildings', *Facilities*, Vol. 10, No. 4, April.
Pepperell, N.T. (1990) *Risk Management for Architects*, RIBA Indemnity Research, London.
Raftery, J. (1993) *Risk Analysis in Project Management*, E & FN Spon, London.
Singleton, W.T. (1987) *Risk and Decisions*, Wiley, New York.

8 | Building performance

OVERVIEW

The built environment should provide users with an essentially democratic setting, enriching their opportunities by maximizing the degree of choice available to them.

Centre for Facilities Management

The design and management of systems of production that help to improve those who work in them, and at the same time more efficiently produce better products, is a task that requires the development of new knowledge, methods and languages. Any improvement in these systems entails some sort of definition and measurement of performance. This has to be the responsibility of a team and what is required are explicit statements of performance requirements and effective performance management and delivery through time within available resources. It is in this context that building performance determines the extent to which facilities either support, or can be adapted to, the changing needs of the occupants of buildings.

The total workplace, in which building occupants perform, includes the social and managerial environment as well as the physical setting for work. Working environments have the potential to contribute to all the proposed elements which are concerned with improving organizational effectiveness. It is founded in the belief that such an environment can maintain commitment amongst the members of the organization, provide communication amongst operating units, project a positive and responsible image, enable change, improve productivity and transform materials and components into value for the customer.

From the point of view of the user, the total workplace includes the immediate and extended physical settings, the processes of creation and management of those settings, and the management style and social climate which support or constrain their use. Such 'healthy' workplaces can support the complete physical, social and mental well-being of occupants. Users are central to healthy buildings – rather than being engineered out, they must be provided with insight, information and influence as their perceptions of control are vital to their feelings of comfort and satisfaction with the workplace. Organizations occupying the built environment must create a democratic setting for work to allow self-determination by users.

The appraisal of building performance comprises three steps – representation, measurement and evaluation. These include the identification of user needs, their conversion into performance requirements of buildings and services, and their development into performance-based specifications to meet them. It is a continuous process of evaluation throughout the formative, preparatory and implementation stages of a cyclical process. Such quantitative and qualitative appraisal and measurement techniques are used in space planning, building management, design-in-use and total cost accounting.

The evaluation of the post-occupancy workability of a building in use poses questions for facilities managers who carry a prime responsibility for ensuring that buildings do work for the organization and for the users. Post-occupancy evaluation tools comprise a battery of techniques including surveys, interviews and walk-throughs, each of which provides essential feedback from building users. They are central to facilities management and demand a basic understanding and ability to apply them.

Keynote paper

Intelligent building performance

Paull Robathan

INTRODUCTION

The overall performance of a building should be assessed by the combined performance of the building as it is affected by the technical capability of the building, the technological environment, the business and its processes and, perhaps most importantly, the individuals involved. Given this, building performance is clearly not static and a knowledge of trends in business and technology are required to be able to predict the changing demands.

From each point of view building performance is perceived as having different goals, but ultimately the effectiveness of the building is measured by its capability to balance the conflicting and contradictory demands placed on it. A key part of the facilities manager's role is to ensure that the building stock available performs suitably for the tasks required from all aspects, both now and in the foreseeable future. However, he or she must look to supporting the organization wherever it exists, in the main office, branches, at home or in hotels, cars or even on the street.

FACILITIES MANAGEMENT IN CONTEXT

Facilities management can be viewed at a number of different levels. At the 'lowest' level – the day-to-day support of operations – a manager of facilities functions can be involved in the maintenance of boilers, provision of coffee machines, mowing the grass and other day-to-day tasks needed to maintain the fabric and services of a static property (maintenance manager role); in organizing photocopying services, controlling security access, the management of ID cards and publishing internal telephone directories, etc. (office manager role); in installing cabling for power, data and telephone services, etc. (communications manager role), and so on. A long list of similar services such as the organization of messengers, microfilming, catering and the car fleet will also be undertaken within the facilities function under the auspices of the facilities manager.

The role, however, does not operate only at this level. The facilities manager also has a key role in the planning for service provision – the planning function – based on business demands. Here the responsibilities include space planning (establishing space standards, making bulk purchases of compatible furniture systems, managing the efficient provision of working space); building projects (preparing the project brief, managing the building team, lease negotiations, acting as the tenant and/or landlord); building management systems; resource management; health and safety; and continuity planning. All these tasks require the effective management of a complex set of interacting services and systems for the good of the business. At this level the facilities manager acts on strategic demands, and develops tactical plans in line with the strategy.

It is at board level – the director of facilities – that the buildings, plant and services of a business need to be managed as assets that generate return on investment. The facilities director is equally as critical to the business as is the human resources or information systems directors. Each in their own way must offer the business competitive advantage through strategic investment. Investing in a building which will be incapable of allowing rapid expansion or contraction of business groups, or one that has services that are under- or over-provided in relation to the business needs over the long term can be a major adverse factor in a firm's competitive ability. Buildings – as property – are assets to be used to the long-term advantage of the business. Indeed it is only when organizations take the facilities director fully into the strategic planning process that the effect of the proactive management of facilities can be appreciated.

MEASURING THE PERFORMANCE OF BUILDINGS

Building performance can be addressed from four different aspects. There is, firstly, the underlying technical capability of the building – perhaps the traditional aspect of building performance. It includes such issues as energy (BREEAM), green buildings and the individual aspects of solar gain, fabric and structure.

Secondly, there is the technological environment, which must present opportunities for flexible location and relocation of computing equipment, telephones, video conferencing equipment and network controllers. It also includes such issues as minimum cost per move/maximum flexibility of location; structured wiring schemes with cable management; flexible power through underfloor bus or in-furniture modules, etc.

Thirdly, business is looking for the support of a rapidly changing working environment. The issues here involve moves from open-plan to cellular divisions, and back, without significant redesign or adjustment of building services (air-handling, power, lighting, communications). Building services must be cost-effective and susceptible to management through business-accessible facilities, such as computer-aided facilities management and building management systems. The response to legislation (e.g. EU Directives, BSI requirements) is another issue at this level.

Lastly, there is the ultimate focus of the individual looking for human scale and the ability to directly control his/her own micro-environment. Opening windows, computer-linked temperature controls, local air movement and quality adjustment, task lighting, access to personal filing, privacy (with a sense of community), limited security restrictions with maximum protection of personal items, comfort facilities (washrooms, meeting areas, vending and dining facilities), interior decor, plants, art, outlook and aspect are all involved.

OVERALL BUILDING PERFORMANCE

Performance is a dynamic phenomenon. A building built before the 1970s performs badly against the norms of the 1980s, though it may perform very well against the norms of the 1990s. As technology changes, and as business processes change, the assessment of a building's performance must also change in step. For example, high quality space with greater emphasis on the 'club' atmosphere and with no fixed persons/desk relationships involves a move to hot desking, teleworking, location-independent working and virtual meetings. When combined with business re-engineering and workflow it presents a challenge to facilities managers.

THE CHANGING NATURE OF WORK

Information technology (IT) and telecommunications are radically altering the nature of work. The technology is brought into every workplace – at ever-decreasing cost and increasing power and compatibility – as computers, facsimile machines, modems and the like. As well as the use of normal voice lines there are other more sophisticated connection services that allow a single message created on a variety of media to be delivered to a range of data terminals, with the added certainty of guaranteed delivery and confirmation of receipt. Digital data manipulation, storage and transmission can create closer networks linking workers.

Cellular telephones, and personal communications networks (PCNs), can liberate the mobile worker from his or her office completely. Adding cellular radio to data products can create a notebook-sized industry-compatible PC with data, fax and voice links; a portable printer completes the truly mobile office. The image, in its many forms (video, colour pictures, graphics), is also a part of sophisticated communication. The integration of these disparate elements into multimedia allows 'virtual reality' in the building marketplace enabling, for example, potential occupiers to 'walk around' inside a building before it has even left the drawing board. Computer-supported collaborative work (CSCW) is the term given to computer and communications integration.

Time and place, the two constraints on world-wide business, are being rapidly eroded. Groupware products can link hundreds of staff in geographically remote locations into electronic conferences, allowing the sharing of information and collaboration on the preparation of consensus documents. By integrating mobile data, voice and groupware with video conferencing, the opportunity exists for 'virtual meetings', where no two participants share the same physical space. Electronic mail and groupware allow participants to be involved in a meeting at separate times.

Workflow automation is the automation of the flow of events making up the series of tasks carried out by people individually or in groups. Each workflow determines what information is required, what task must be done, who should perform the task, what deadlines should be applied and under what rules or policies the activity should be performed. Workflow is an inherent part of business process re-engineering, which itself is based on total quality management initiatives.

THE ROLE OF BUILDINGS IN BUSINESS

The traditional view of firms is changing. Location need no longer be the prime motivation in choice of property, and the premises a firm occupies should no longer restrain the implementation of new technology. There are two views of services provided for an organization: building-centred and organization-centred.

Building-centred services

Buildings are changed from inflexible, high cost, low functionality spaces into efficient integrated communications and control environments. In the traditional building the power supplies, air-conditioning systems, lighting, external fabric, security systems, communications and computers all operate independently, and the humans inside the building spend a proportion of their time struggling to satisfy the conflicting demands. But add a comprehensive and integrated cabling scheme, and interrelate the various subsystems through a single control framework, and the building can begin to respond to its environment in an effective manner. The benefits of such so-called intelligent buildings are integration for control and efficiency through flexibility.

To achieve these benefits requires the application of a holistic approach to building design and construction. The framework within which decisions are taken must be broad and all-embracing, with the aim of generating an infrastructure to underpin the changing demands of the occupants. The various disciplines involved in building services design have to coordinate their actions such that the resulting design incorporates a unified view of the building as a functioning unit, rather than a shell with many independent, contradictory and inefficient services.

Whatever the cost at inception, the lifetime costs of building management are potentially much lower with the incorporation of intelligence. The flexibility of use of the building space is enhanced, the occupants more productive, and the rental return higher. Particular examples are:

- **Building energy management systems (BEMS)**. This is the integrated monitoring and control of the energy systems in a building designed to ensure building services plant works to optimal efficiency. It is based on a network of control outstations linked to one or more computers operating a set of control algorithms.
- **Integrated air-handling, cabling and space management**. This is the coordinated support of flexible space management through the installation of an integrated under-floor air delivery system, removing the need for ceiling-based equipment and cabling.
- **Computer-aided facilities management**. This is the recording, management and control of all aspects of building design, furniture location, cables, staff–equipment relationships and many other services in a building or portfolio of buildings. Such systems can be programmed to produce work dockets of planned maintenance and to generate orders for spare parts as locally stored spares are used up.

The impact buildings have on their environment is a growing concern. The use of environmentally friendly substances, the elimination of CFCs, using only cost-effectively reproducible resources (softwoods instead of hardwoods) and the minimizing of noise, gas emissions and visual detraction are major factors in the design or modification of any building. Within the building furniture systems must conform with the EEC VDU Directive, as incorporated into the health and safety regulations from 1993.

Organization-centred services

The facilities manager must support all elements of the organization wherever they happen to be. It should be noted that the information, telecommunications and ultimate business processes that create a business are not building-centred. The aspect of security, particularly of data, is becoming critical, with mobile users networking into sensitive files, and being able to mail those files from anywhere to anybody (inside or outside the company) using the public mail services – access control looks like locking the stable door after the horse has bolted!

The dramatic changes in work, and the close links between work and buildings, have created a new class of supplier. Systems integration can be a significant force in the provision of building/organizational solutions for business and IT suppliers have established new entities specifically to address this market.

INTELLIGENT BUILDINGS

The intelligent building has been discussed, defined and investigated for several years. Intelligence is a characteristic that appears incongruous when applied to any inanimate object, and in particular to a building. However, dictionary definitions of intelligence – 'quick to understand, sensible' and 'giving information, communication' – provide a clear indication of why the description is relevant to the needs of businesses. Intelligent buildings straddle two vastly different, but essential, demands made by organizations. Buildings must respond to changing demands effectively and quickly, and must support a high level of inter-working between workers.

Intelligent buildings can only be considered successful by being appropriate to the demands made on them and these vary over time. In the 1990s the demands on offices are for adaptable, low cost space capable of delivering high levels of performance to individuals operating in a networked firm. Personal computers and telephones connect people together in a collection of operating units which present themselves as an integrated business. Are intelligent buildings the answer?

The organizational and technical factors that seem to have created the demand structure for intelligent buildings, together with the role of the facilities manager in satisfying the often conflicting demands of individuals, firms and properties, are described below.

Benefits from intelligent buildings

The occupant of an intelligent building can receive many benefits from the application of intelligence to the management of his or her workplace. These benefits are both physical, through efficient and responsive lighting, heating, acoustical and security services, and also mental through integrated and efficient communications between individuals and between people and computers.

The managers of an intelligent building receive benefits in the quality of control they are afforded over services and systems, and the feedback provided by the infrastructure itself. When facilities managers can set parameters for building-wide systems, and retain control over the cost/performance of these systems, then management control becomes a reality, whereas in the conventional building facilities management is a fire-fighting exercise in response to crises without control.

These managers also gain when any change of personnel or services location takes place. Rather than having to lay new cables, disrupt services, reassess lighting, power and air-conditioning on an *ad hoc* basis, managers can rest assured that the services within the building will cope with the basic needs, and leave the facilities managers to control the physical equipment and staff moves that are necessary. Plans can be set in advance, costed, scheduled and adhered to. The managers also gain by getting the building management system to prepare planned preventative maintenance schedules automatically, together with work orders.

The owners of an intelligent building receive a direct cost benefit from a reduction in lifetime costs for the management and operation of the building. The initial cost is more predictable, due to a common set of consistent services and facilities specifications being available. It is no longer necessary to make all the short-term decisions about computers, location of staff and use to be made of each area. Buildings can be built to a size and design in keeping with long-term demands while being internally responsive to virtually all demands.

The planners, architects and professional teams involved in creating the building can have available, through the facilities manager, a comprehensive offering capable of providing the integrated services, facilities and management features essential in the intelligent building. The advent of supplier-independent facilities complying to international standards will allow designs to be substantially completed, building use parameters to be established, and the ultimate customer to have a clear notion of the cost/performance features of the building even before work begins on the site.

The challenge of intelligent building performance

The opportunity to implement new working methods, and the incorporation of intelligent building techniques into premises, presents firms with a series of performance challenges. Designers of buildings and of working environments must anticipate the trends that will affect buildings during their expected life. The influences abound – lower power needs, lower air-conditioning implications, significantly less wiring, more mobility of the individual, areas for impromptu meetings, workplaces that are shared on an as-needed basis with little territorial restriction, and instant country-wide (and global) communication for both speech and data.

Buildings with 600 cm false floors, 4 m slab to slab and cavernous risers, can either end up under-used or be modified to let the outside world in again with opening windows, extensive revision of air-handling plant, and the creation of deliberate open circulation rather than delineated office boundaries. The spiral of demand from computer to cabling to power to air-conditioning to special building construction will be broken, and building design will be able to free itself of technological restraint and return to the task of making people more comfortable and effective.

CONCLUSIONS

Such 'intelligent workplaces need to support the complete physical, social and mental well-being of occupants. Users are central to the intelligent buildings – rather than being engineered out, they must be provided with insight, information and influence as their perceptions of control are vital to their feelings of comfort and satisfaction.

This is the goal of intelligent building performance – creating and sustaining an environment which maximizes the efficiency of the building's occupants while enabling effective management of resources at minimum lifetime costs. The definition does not mention technology, nor is it tied to specific design or implementation. Intelligent building – the application of intelligent techniques – is intended to match the needs of the occupier at any time with appropriate, cost-effective support and save money over the lifetime of a property investment.

SUMMARY

All buildings require, throughout their life, a level of performance and a standard of management that can provide and sustain conditions suitable for the well-being of their users. In the same way that facilities management responsibilities in an organization can be seen at a number of levels, so too can building performance. The built workplace performs in a way which is sophisticated in terms of technology, as evidenced by the 'intelligent buildings' scene, and complex in terms of patterns of organizational change.

Building performance for facilities managers is taken to refer to the extent to which a building responds to the needs of its users. Intelligent buildings, once a byword for all things technological in buildings, is also seen from this point of view. Similarly the 'responsible workplace' may be described as a workplace which is responsive to users, adaptable to organizational change and customizable rather than being custom made. The total workplace describes not only the physical facilities such as furniture, workstations and partitions, but also takes into account the whole network of social, organizational and design elements that constitute the context of employees' working lives. It is of particular significance in relation to the 'sick-building syndrome', and other related problems, as their causes are a complex mixture of physical and psychological factors.

From the point of view of the user, the total workplace includes the immediate and extended physical settings, the processes of creation and management of those settings, and the management style and social climate which support or constrain the rich use of them. These concepts seek to change people's attitudes to them, to see them as more than their own personal workstation, and to perceive that their workplace includes all aspects of their working environment.

BIBLIOGRAPHY

Essential reading

Becker, F. (1990) *The Total Workplace: Facilities Management and the Elastic Organization*, Van Nostrand Reinhold, New York.

Duffy, F. (1990) *Measuring Building Performance*, Facilities Management International Conference Proceedings, Centre for Facilities Management, University of Strathclyde.

Editorial (1990) *Building Performance*, Facilities Management International Conference Proceedings, Centre for Facilities Management, University of Strathclyde.

Zeisel, J. (1981) *Inquiry by Design: Tools for Environmental Behaviour Research*, Cambridge University Press, Cambridge.

Recommended reading

Becker, F. (1981) *Workspaces: Creating Environments in Organizations*, Praeger, New York.

Duffy, F. (1992) *The Changing Workplace*, Phaidon, London.

Duffy, F., Laing, A. and Crisp, V. (1993) *Responsible Workplace*, Butterworth Architecture, Oxford.

Preiser, W.F.G., Rabinowitz, H.Z., and White, E.T. (1989) *Post Occupancy Evaluation*, Van Nostrand Reinhold, New York.

Ruck, N.C. (ed.) (1989) *Building Design and Human Performance*, Van Nostrand Reinhold, New York.

Sundstrom, E. (1986) *Work Places: The Psychology of the Physical Environment in Offices and Factories*, Cambridge University Press, Cambridge.

Further reading

Bechtel, R., Marans, R. and Michelson, W. (1987) *Methods in Environmental and Behavioural Research*, Van Nostrand Reinhold, New York.

Brill, M., Margulis, S. and Konar, E. (1984) *Using Office Design to Increase Productivity*, BOSTI, Buffalo, NY.

Davis, G., Becker, F., Duffy, F. and Sims, W. (1984) *ORBIT 2.1*, Harbinger Group, Newark, CT.

Duffy, F. (1989) *The Changing City*, Bulstrode Press, London.

Joroff, M., Louargand, M., Lambert, S. and Becker, F. (1994) *Strategic Management of the Fifth Resource: Corporate Real Estate*, Industrial Development Research Foundation, GA.

Kernohan, D., Gray, J., Daish, J. with Joiner, D. (1992) *User Participation in Building Design and Management*, Butterworth, London.

Oxford Brookes University, School of Estate Management, and University of Reading, Department of Land Management and Development (1993) *Property Management Performance Monitoring*, GTI, Oxford.

Powell, J.A., Cooper, I. and Lera, S. (1984) *Designing for Building Utilisation*, E & FN Spon, London.

Steele, F.I. (1973) *Physical Settings and Organizational Development*, Addison-Wesley, Reading, MA.

Wineman, J. (1986) *Behavioural Issues in Office Design*, Van Nostrand Reinhold, New York.

<table>
<tr><td>9</td><td># Environmental management</td></tr>
</table>

9 Environmental management

OVERVIEW

The freedom to act is not a licence to abuse.　　　Tom Cannon

As innovation and enterprise are the key to business success, a challenging and changing scenario creates new opportunities and imposes new responsibilities on firms and their managers. The term 'green' has come to describe an approach that seeks to respect the environmental consequences of any action, with a view to conserving the ecology of the planet as a whole. Organizations must work to demonstrate sound environmental performance. They will see 'green' as a key marketing tool, and make public statements of policy and performance to establish an organization's environmental credentials.

Surveys have shown that a majority of companies in the United Kingdom and Europe view an effective corporate response to green issues as a key to success. Surveys of company directors also reveal many conflicts of interest in dealing with environmental issues among UK businesses, especially those in the service sector. European companies often have a formal environmental policy, typically entailing a commitment to limiting environmental impact to a practical minimum.

Although companies should undertake full environmental audits in the form of systematic, objective and independent analyses of the environmental impact of business, many may only undertake reviews of the key environmental issues as it affects them. A framework therefore needs to be developed for the definition of corporate responsibilities to the environment under which the factors in the development of organizational environmental policy can be considered.

The recognition of environmental issues at individual and corporate level requires the development of an understanding of the techniques and systems for environmental and energy management. The use of these techniques and the application of such technologies will determine and control the use of energy and environmental impact of a range of buildings in both the public and private sectors.

Perhaps, not surprisingly, there is often a tendency to concentrate more on the physical rather than the social and symbolic aspects of the environment of buildings and on their human occupancy. There is also a tendency to focus more on delivery and product than on the processes of sustaining quality for the occupants. There is some attempt to develop 'packages' that sell the concept of an environmentally conscious, ecological and user-friendly approach, but is it comprehensive enough?

The need is to redefine environmental quality and change both the definition and delivery of what is meant by comfort. The traditional approach to specifying 'comfort' simply in terms of psycho-physiological criteria is just too limited. Instead the need is to move towards criteria for environmental quality which can encompass not just economic, social and psychological factors but ecological and political ones as well.

Achieving and sustaining environmental quality – from the micro to the macro level – depends just as much on the ways in which the environment is managed as on the choices made about the materials and methods used in building construction. It also depends on the plant, services and energy sources employed to deliver specific aspects of internal conditions.

Keynote paper

Emerging issues in environmental management

Ian Cooper

INTRODUCTION

As part of their support to their organization's effectiveness and well-being, facilities managers need to address issues of environmental management. Those facilities management activities such as purchasing, building procurement, facilities and energy management, and fleet transport management can all be seen simply as specialist contributions against the overriding significance of environmental management – which needs a coherent conceptual framework to tackle it responsibly. For this it is necessary to examine the scope of environmental management and to identify its importance for facilities managers. One of the first issues which has to be faced is how an organization should formulate and implement an environmental policy and what role facilities managers should have in this process.

ENVIRONMENTAL MANAGEMENT DEFINED

In the terms of British Standard BS7750: 1992 Specification of Environmental Management Systems produced by the BSI in 1992, environmental management can usefully be described as the processes and practices introduced by an organization for reducing, eliminating and, ideally, preventing negative environmental impacts arising from its undertakings. Expressed more positively, the key goal here is the achievement of environmental quality, whether measured narrowly in terms of the working conditions an organization provides for its staff or, more broadly, in terms of the overall impact of all its operations and activities on the local, regional and global environment.

In order to protect the environment at all three of these levels, and to safeguard that its personnel occupy healthy buildings, an organization needs to ensure that:

- its operations and activities comply with environmental legislation;
- its products or services are procured, produced, packaged, delivered and used, and ultimately disposed of, in environmentally appropriate ways;
- its expenditure (both in terms of staffing and resources) on environmental protection is timely and effective;
- its strategic planning for future investment and growth reflects market needs concerning the environment.

To provide this level of environmental protection, an organization needs to know and understand the significance of the environmental effects of three things, namely:

- the 'upstream' production of the resources and materials it uses;
- its own activities and operations;
- the 'downstream' use and/or disposal of its products and services.

This specification defines environmental management extremely broadly, as an open-ended set of responsibilities which stretch from 'cradle to grave', or from 'womb to tomb' so to speak. It also embraces the Friends of the Earth maxim – 'think globally, act locally.' In this sense, the Standard formalizes and institutionalizes what were previously only fringe pressure-group concerns, taking them into the mainstream of organizational management in the UK.

The British Standard recommends that, to achieve this, an organization has to make sure not only that all the technical factors affecting its environmental impact are under control, but that all its administrative and human factors are effectively managed as well. And, to accomplish all of this, it proposes that an organization should:

- identify the key elements affecting its environmental performance;
- develop an integrated management system for regulating them;
- define a policy containing a set of objectives for managing these effectively;
- introduce a method for reviewing how effectively the policy and objectives are being met.

Significantly, the Standard takes it as axiomatic that any such policy will establish environmental goals which are more stringent than current regulation or legislative requirements, although these are presented as forming the minimum baseline.

Operation of the environmental management system which an organization introduces is then to be internally audited and evaluated, on a regular basis, to assess the effectiveness of the system in achieving stated objectives. However, ultimate responsibility for, and commitment to, environmental policy, it states, has to belong to the highest level of management.

FACILITIES MANAGEMENT IMPLICATIONS

The open-ended nature of the specification of environmental management in the British Standard, along with its insistence that compliance with existing regulations and legislative requirements should form only the minimum of an organization's environmental objectives, are both pregnant with significance for how, in future, facilities managers will define and seek to discharge their responsibilities. These characteristics of the Standard signal two quite clear movements, both of which will move facilities management from left to right in Figure 9.1.

First, they mean a movement away from mandatory compliance towards voluntary responses, from simply complying with existing legislation towards organizations developing their own proactive stances towards reducing their environmental impact. Second, and in parallel, they mean movement:

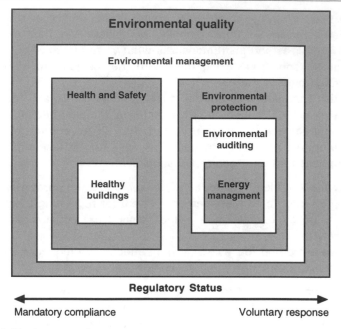

Figure 9.1 Environmental management components

- away from primarily considering the internal environment in terms of meeting health and safety legislation;
- through a concern for the delivery of healthy buildings;
- broadening out via the introduction of tools for auditing both the internal and external environment of buildings;
- towards the strategic management of facilities, both in terms of their impact on the internal and external environments and vice versa.

AUDITING ENVIRONMENTAL IMPACT

One pertinent illustration of how developments on this front are shaping up is BREEAM, the Building Research Establishment's Environmental Assessment Method for buildings. This was originally a labelling system created solely as a design aid, as a means of providing guidance to architects and engineers involved in producing new buildings on how to reduce their environmental impact. A number of versions of BREEAM now exist, including those for the design of new offices, supermarkets and houses.

Another version of BREEAM is targeted on the environmental labelling of existing offices. This version also responds to two perceived shortcomings focused on new buildings. First, each year new additions are rarely added to the building stock at a rate much above 1 per cent, even outside a recession. So any advice focused solely on new buildings will take a long time to have a significant effect on the environmental performance of the built environment. Secondly, the environmental performance of buildings is not just a design issue – it also depends upon how buildings are managed and used.

This version of BREEAM is being presented, quite explicitly, as a management aid since, once buildings are constructed, they have to be managed. Good management reduces environmental impact, sometimes dramatically, a point put more fully by Doggart in *Management – A Key Environmental Issue*, a paper presented at a Buildings and the Environment Conference at Queens' College, Cambridge, organized in 1992 by the University of British Columbia. Accordingly, the guidance focuses directly on management issues by giving credits, for instance, for possessing:

- environmental policies and actions which show the commitment of top management methods for achieving energy policies (such as audits, targets, maintenance schedules and resources);
- policies on waste management and recycling;
- policies for the management of sick building syndrome.

The value of auditing facilities by use of a labelling scheme such as BREEAM is that it provides an external datum against which to measure performance. The disadvantage is that this datum is not geared to an organization's specific needs and requirements. These can be met, however, by internally defined auditing procedures.

This is the direction signalled by the Building Services Research and Information Association's *Environmental Code of Practice* published in 1992. In its section on facilities management, the Code proposes that, in the first instance, emphasis should be:

To establish a management and financial strategy for improvement based on lifecycle costs and environmental impact. It is important that the initial commitment is made at a high level to review the company's facilities management operation and maintenance and to establish funding options with regard to progress. Environmental costs and benefits should be considered alongside lifecycle costs.

The Code recommends that such an environmental strategy should be developed through a planned and balanced raising of management awareness over time. And it proposes that this should be done by adopting a diagnostic tool produced for the Building Research Energy Conservation Support Unit. As described in General Information Report 10 published by BRECSU in 1992, this was a matrix originally devised for identifying a company's organizational profile in relation to energy management as shown in Figure 9.2.

Level	Energy policy	Organizing	Relating to users	Information systems	Marketing	Investment
4	Energy policy, action plan and regular review have commitment of top management as part of an environmental strategy	Energy management fully integrated into management structure. Clear delegation of responsibility for energy consumption	Formal and informal channels of communication regularly exploited by energy manager and energy staff at all levels	Comprehensive system sets targets, monitors consumption, identifies faults, quantifies savings and provides budget tracking	Marketing the value of energy efficiency and the performance of energy management both within the organization and outside it	Positive discrimination with detailed investment approval to exploit all new-build and refurbishment opportunities
3	Formal energy policy, but no active commitment from top management	Energy manager accountable to energy committee representing all users, chaired by a member of the governing body	Energy committee used as a main channel together with direct contact with major users	M&T reports for individual premises based on sub-metering, but savings not reported effectively to users	Programme of staff awareness and regular publicity campaigns	Some payback criteria employed as for all other investment
2	Unadopted energy policy set by energy manager or senior departmental manager	Energy manager in post, reporting to ad hoc committee, but line management and authority are unclear	Contact with major users through ad hoc committee chaired by senior departmental manager	Monitoring and targeting reports based on supply meter data. Energy unit has ad hoc involvement in budget setting	Some ad hoc staff awareness training	Investment using short term payback criteria only
1	An unwritten set of guidelines	Energy management the part-time responsibility of someone with only limited authority or influence	Informal contacts between engineer and a few users	Cost reporting based on invoice data. Engineer compiles reports for internal use within technical department	Informal contacts used to promote energy efficiency	Only low-cost measures taken
0	No explicit policy	No energy management or any formal delegation of responsibility for energy consumption	No contact with users	No information system. No accounting for energy consumption	No promotion of energy efficiency	No investment in increasing energy in premises

Figure 9.2 Energy management matrix (BRECSU, 1993)

FORMULATING STRATEGY

Ideally, an organization's environmental strategy should be formulated and implemented from the top down, and it should be driven throughout the whole organization by the explicit commitment and encouragement of top management. This is best done through a formal, written environmental strategy. It can then operate both as a public expression of an organization's commitment to action and as a working document for guiding that action over time. Where an organization is already committed to developing a formal environmental strategy, then the mechanisms open to facilities managers for making their own contribution to formulating and implementing environmental policy should be evident. In these advantageous circumstances, they simply need to develop a coherent overview of how their own function impinges on their own environmental performance.

Figure 9.3 contains a simplified conceptual map illustrating how facilities-related concerns fit within an organization's overall approach to environmental management.

Once facilities managers are clear about their own circumstances, then they need to identify how, when and where within the formal and informal structure of their organization they can most effectively contribute to formulating and implementing policy.

A facilities manager, however, may not find him/herself in this fortunate position. Their organization may not yet have committed itself to an environmental strategy. But such facilities managers may still wish

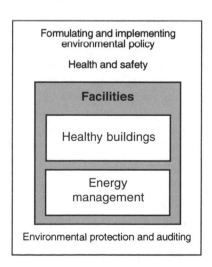

Figure 9.3 Conceptual map

to reduce the environmental impact of their own activities and operations. In these circumstances, they first need to diagnose – using a tool like the BRECSU energy management matrix – where their organization currently stands in terms of good practice in the environmental management of its facilities, and where its present strengths and weaknesses lie.

Once they have discovered where progress needs to be made they can then identify current opportunities for, as well as obstacles to, improving the environmental impact of these facilities management practices. This approach can be used to document how existing practices appear from where they stand within their organization. Once this has been done, they can begin negotiating with their line managers to raise the awareness of senior managers about what needs to be done.

THE STRATEGIC OPPORTUNITY

As BS7750 illustrates, reducing an organization's environmental impact is a strategic issue which not only reaches across each constituent part of its own operations and activities, but also extends to include those of its suppliers and those who use its goods and services.

Facilities management also needs to be seen as a strategic concern which, at its most mature, permeates right through an organization's non-core support structures and operations.

The rise of environmental management represents an opportunity for facilities managers to illustrate the pervasiveness of their contribution to their organization's effectiveness and well-being. They can use this opportunity to demonstrate that, at its broadest, facilities management extends beyond technical considerations involving the provision and maintenance of premises and accommodation.

By participating in formulating and implementing policies for reducing their organization's environmental impact, facilities managers can demonstrate the value of redefining their role as one which underpins mainstream organizational and business needs, rather than just as functional managers responsible for a non-core activity based on a marginal technical expertise.

SUMMARY

Environmental management is the effective management of an organization's activities, in order to ensure that they meet defined environmental objectives, and the process by which an organization sustains a healthy, safe and ecologically sound environment to meet its strategic objectives. It is also a system whereby competent and committed employers, with the participation of their employees, identify health and safety threats and adverse environmental effects and select, implement and monitor measures within a framework of law and standards drawn up with the involvement of all parties.

Effective management of an organization's activities is required in order to ensure that they meet defined environmental objectives. These objectives should be embodied in a comprehensive environmental management system consistent with total quality systems.

Organizations that manage their response effectively and reposition themselves to suit are capable of building more secure and more prosperous positions in the marketplace. Successful compliance with the terms of a business contract to sustain the environment will generate profits in the long term and add value to the enterprise as well as the community. Facilities managers need to understand how the growing emphasis on reducing their organization's environmental impact will move them into areas of building health and environmental auditing and affect the way they discharge their duties.

Most firms operate in an urban or built environment. The buildings in which people live and work have an immediate effect on their lives and lifestyles. The quality of the built environment is perhaps the most tangible expression of the contribution that business makes to the health of the society in which it operates.

BIBLIOGRAPHY

Essential reading

Cannon, T. (1992) *Corporate Responsibility*, Financial Times/Pitman, London.
Carmichael, S. and Drummond, J. (1989) *Good Business: A Guide to Corporate Responsibility and Business Ethics*, Business Books, London.
Carson, R. (1962) *Silent Spring*, Penguin, Harmondsworth.
Copper, D. and Palmer, J. (eds) (1990) *The Environment in Question*, Routledge, London.

Recommended reading

Doggart, J (1992) *Management – A Key Environmental Issue*, paper presented to a Buildings and the Environment Conference at Queens' College, Cambridge, organized by the University of British Columbia.
Energy Efficiency Office, DoE (1993) *Making a Corporate Commitment*, HMSO, London.

Mintel Special Report (1989) *The Green Consumer*, Mintel, London.

Pearce, D., Markadya, A. and Barbier, D. (1989) *Blueprint for a Green Economy*, Earthscan, London.

Touche Ross Management Consultants (1990) *Head in the Clouds or Head in the Sand: UK Managers' Attitudes to Environmental Issues*, Touche Ross, London.

Vischer, J.C. (1989) *Environmental Quality in Offices*, Van Nostrand Reinhold, New York.

Further reading

Bechtel, R., Marans, R. and Michelson, W. (1987) *Methods in Environmental and Behavioral Research*, Van Nostrand Reinhold, New York.

Becker, F. (1981) *Workspaces: Creating Environments in Organizations*, Praeger, New York.

British Standards Institution (1992) *Specification of Environmental Management Systems*, BSI, London.

Building Research Establishment Conservation Support Unit (1992) *Reviewing Energy Management in Private Companies*, General Information Report 10, BRECSU, Watford.

Building Services Research and Information Association (1992) *Environmental Code of Practice*, BSRIA, Bracknell.

Criswell, J.W. (1989) *Planned Maintenance for Productivity and Energy Conservation*, Fairmont Press, New York.

Department for the Environment (1991) *Environmental Action Guide for Building and Purchasing Managers*, HMSO, London.

Halliday, S.P. (1994) *Environmental Code of Practice for Buildings and Their Services*, BSRIA, Bracknell.

Prieser, W., Vischer, J.C. and White, E. (1991) *Design Intervention: Toward a More Humane Architecture*, Van Nostrand Reinhold, New York.

Prior, J.J. and Bartlett, P.B. (1995) *Environmental Standard: Homes for a Greener World*, Building Research Establishment Report No. 278, Watford.

Information management

OVERVIEW

My optimism comes from the empowering nature of being digital.
The access, the mobility and the ability to effect change are what
will make the future so different from the present.

Nicholas Negroponte

Information is a key resource and it is the ever-increasing power of information technology (IT) which lies at the heart of most businesses. Indeed, progressive business leaders look to IT to help integrate their changing organizations. This integration, given an effective operation, provides a context within which to consider the planning and provision of the IT service. Without the right people, organizations cannot expect to deliver the levels of competence, problem-solving and courtesy demanded of them. The complexity of modern business relies on the skills and knowledge of its employees who, of course, are also its customers. Increasingly this IT-based integration embraces these customers who are thus better informed, have more choice and therefore have more power. Service organizations are an example of those who particularly depend on having the right information in the right place at the right time.

Communication and decision-making are dependent on information. In its electronic form it increases the capability of organizations and can revolutionize the ways in which its activities are arranged through space and time. For example, the importance of information to organizations can lead to a fundamental rethink of the relationship of buildings and computers and the information systems they contain. In effect a building can be an information infrastructure created by organizational designers with an enclosure wrapped around it. The scope of facilities management is such that it can easily embrace this kind of development, and indeed might also cover specific responsibility for managing the information itself.

Information management is the process by which quality information is generated, structured and communicated to support its application in decision-making. Organizations must develop a coherent framework that includes the information required for planning, providing and managing facilities. Specific facilities needs will be embodied in a facilities information management system (FIMS), consistent with the organization's overall systems.

Facilities management is defined in the IT industry as 'the provision of the management, operation and support of an organization's computers and/or networks by an external source at agreed service levels'. The service will generally be provided for a set time at an agreed cost and requires a framework around which to develop a facilities information management strategy. This allows for the identification of organizational needs and objectives, and also information for decision support systems at strategic, tactical and operational levels. Organizational strategies for information structuring and database integration can also be determined. It enables the process to be implemented and provides criteria for systems selection with an overview of systems availability.

Keynote paper

Developing facilities information management systems

Ameen Joudah

INTRODUCTION

The automation of facilities management functions can significantly improve their effectiveness in organizations. Adopting, developing and maintaining an appropriate level of knowledge and skill in information management and technology is crucial therefore to the effectiveness of facilities management teams. As volume and complexity increase and information gets difficult to handle, computers are becoming more widely available and affordable, and are being readily adopted for this purpose. There is also a large number and variety of relevant (if not always suitable) computer-aided facilities management (CAFM) systems available to 'aid' FM processes and operations.

This chapter defines the meaning and strategic essence of CAFM and of facilities information management systems (FIMS), outlines a strategic FIMS framework and its parts, and establishes key criteria for appraisal. However, many users may still express dissatisfaction or confusion, and many vendors have difficulty in competing to produce systems to suit user requirements, so the chapter concludes by briefly addressing the key issues that can rectify such concerns.

THE ESSENCE OF FACILITIES MANAGEMENT AND THE BASIS FOR FIMS DEVELOPMENT

What is facilities management?

Many current definitions of FM give emphasis to technical issues at an operational level. In this the strategic significance of facilities management to business organizations is largely overlooked. A more holistic and strategic understanding of FM is necessary in order to achieve optimum organizational effectiveness.

A fuller definition would be 'the planning, design, procurement and maintenance of all property assets and their associated support and customer services to achieve and sustain optimum environmental quality and efficiency to achieve best value for investment within appropriate resources within the law'. This links two main functions: the management of physical assets (estates) and the provision of support services (hotel services). Facilities management is concerned with managing quality, value and risk associated with the occupancy of buildings and the delivery of customer services. This is described more fully by Keith Alexander in his 1992 paper 'Quality managed facility' (*Facilities*, Vol. 10, No. 2).

Facilities management is not just dealing with records and inventories of facilities and their maintenance schedules at an operational level – as some still unfortunately believe. It is a total set of processes that operates at three levels: strategic – where key planning decisions are made; tactical – where analysis and design processes take place; and operational – where implementation and day-to-day running of facilities processes are handled. These decision-making processes are strategic, complex, cover many areas and need to be subject to multiple criteria analysis. The essence of FM is making decisions for organizations and this should be the basis for FIMS and CAFM development.

What is computer-aided facilities management?

CAFM can be defined as the use of automated 'tools' and procedures which increase productivity and efficiency in facilities management. FIMS reflects the setting up of strategies and frameworks, and the tools used, for managing information at strategic and tactical levels. CAFM should be seen as a subsection of FIMS which deals with the 'computerized' tools or systems through which these information management functions are processed. It is mostly handled and managed at an operational level, but is a consequence of strategic planning, evaluation and implementation.

FACILITIES INFORMATION MANAGEMENT SYSTEMS

The information concerned with FM processes and functions is considerable. Handling it is complex and the way data is structured, collected, collated, distributed, presented and updated determines whether this data is 'informative' and suitable for the process of making various decisions. Of course, information is itself a facility to be managed and a tool by which FM is applied. In other words, it is through effective use of information that FM decisions are made to achieve quality management of facilities. At the same time, as information is an organizational asset, FM concepts (quality, value and risk management) should also be applied to managing it. This should ensure quality managed information.

A well developed information management system ensures that quality records are available, that the decision-making process can be traced, and feedback and feedforward mechanisms are in place to ensure effective communication amongst the facilities team. Furthermore, it should convey messages in 'informative' ways through user-friendly interfaces, using 'responsive' software systems to user requirements. With the proper structuring and interfacing of applications to suit different FM functions, this information management system is a facilities information management system (FIMS). Such a system can be used to manage the following categories of information:

- **physical resources** – site, infrastructure (utility systems), buildings, resources/assets within buildings;
- **support services** – administration, property maintenance, security, health and safety, catering, car fleet management, contract control, cleaning, etc.;
- **human resources** – personnel management, contracts, recruitment, training, etc.;
- **business information** – process documentation, contractual, policies, financial, etc.

WHY DEVELOP A FIMS?

Information should be recognized as power and managed as a key resource with direct implications for business and operational efficiency. It should enable better informed decision-making at strategic, tactical and operational levels. For example, FIMS enables effective control and coordination of technical functions/activities. Appropriate implementation would also increase the return on investment in these systems and procedures. Another key advantage can be to minimize exposure to risk to an organization through loss or the mis-handling of information and databases. Hamer, in his 'Getting a jump on CAFM' article in the January/February 1991 edition of the IFMA *Facility Management Journal*, and Stout, in his paper *The Financial Impact of Computer-Assisted Facilities Management Systems* presented to the Glasgow EuroFM Conference in April 1990, both put this more fully. An appropriate FIMS should be developed not only to create a flexible and reliable base for future activity, but also, if necessary, to rectify any concerns of the kind referred to below.

FIMS DEVELOPMENT CONSIDERATIONS

To develop a FIMS, critical consideration should be given to the following strategic issues. Firstly, the nature and structure of the organization and its business should be studied in order to identify patterns of business and management behaviour and to permit strategies for the effective marketing of the organization and its products. Secondly, project types and their procedures should be identified in order to develop responses to information flow patterns required for the decision-making processes of projects. Thirdly, a critical analysis of 'customer' (internal and external) information requirements should be carried out in order to give responsive and compatible systems. Fourthly, the human resource implications should be investigated in order to assess the availability, skills and training requirements. Budgetary implications also need to be studied to identify the financial feasibility of initiating and maintaining a FIMS system. Finally, other existing resources, such as computing systems, should be optimized.

DEFINING ORGANIZATIONAL REQUIREMENTS

The first step in establishing a FIMS is to identify 'business', 'functional' and 'operational' requirements at all levels. These include those for the facilities management department, whether for a comprehensive and integrated framework, or for a few discrete applications. Requirements stemming from inter-organizational activity also need to be identified. Organizations should conduct a FIMS audit to identify: Where are they now? Where do they want to be? Where are/should they be going? How will they get there? The identification of organizational needs forms half of the basis for analysing and implementing a FIMS. The second half is the development of 'performance indicators' (related to quality assurance, value added and risk consciousness) of such FIMS.

FIMS COMPONENTS

A FIMS should have the following components:

- objectives;
- a strategy;
- a framework;
- constituent applications and systems;
- systems management;
- procedures;
- operations and maintenance;
- evaluation, selection and implementation.

DEFINING OBJECTIVES FOR A FIMS DEVELOPMENT

Key objectives for a FIMS should include treating information, and managing it, as a vital business resource. This gives backbone to the viability of the business operation, and supports decision-making and operational effectiveness. Providing quality information, e.g. the right information in the right format at the right place (source and destination), is another essential objective. Minimizing the risk of information loss and mis-handling is yet another. These objectives should be developed as part of the support for the FM department and the overall organizational objectives. Examples of such objectives include using the information and systems to make better financial analyses, to improve customer interface, to improve staff/process productivity, and to achieve proper control of contractors and their operations. This is explained at greater length in the paper 'Facilities information management systems' in *Facilities Management '93*, published by Hastings Hilton, London, in 1993.

DEVELOPMENT OF A FIMS STRATEGY

A FIMS strategy should fulfil existing (short-term strategy) and future (long-term strategy) objectives and functional requirements. It should also be regularly updated to manage change.

Considerations of available resources – physical, financial and human – would give indications as to what could/should be done in-house and/or contracted out. Human resources particularly require close attention as the availability of required skills and training needs can constitute a hidden cost for a system. An analysis of resources would define a realistic measure for the scale of implementation, though with FIMS it is advisable to 'plan big' but 'implement realistically'.

Technically any strategy should allow for maximum capability and flexibility. This may affect FM functions and procedures. The strategy should also have a suitable system–user interface allowing for maximum user-friendliness irrespective of the sophistication of the technology.

DEVELOPMENT OF A FIMS FRAMEWORK

Appropriate development of a framework for FIMS would involve an analysis of which applications to automate (computerize?), both now and in the future. This analysis should cover aspects such as the interrelationships between these application areas, how to phase them, and their data volume, type, format and interface requirements for data exchange and systems management purposes. Other crucial systems management issues include data security and access to information to minimize exposure to risks, and the systems–user interface to maximize the use value and return on investment of these systems. In addition a critical assessment is required of available information systems and CAFM systems, both within the organization and on the open market. Sophistication (responsiveness to applications) and simplicity (ease of use with minimum training and support) are important criteria for effective evaluation and selection.

A MODEL FOR A FIMS FRAMEWORK

Based on the concepts outlined above, a model for a FIMS framework could be based on two main functional categories common to many organizations as follows.

Office automation functions

This includes most of the organizational administration, finance and personnel office functions as well as FM processes/projects functions, handled separately or by integration. Such functions are 'document-based', where information is grouped, created, handled, edited, updated, stored and presented in a document (report) form. The contents of such documents are classed as 'alphanumeric', i.e. textual, with some graphs and diagrams. A FIMS framework for this category should comprise three main modules, namely:

- document preparation;
- document presentation; and
- document management.

FM process/projects functions

According to the nature of these FM processes/projects and the type/ format of information handled and managed, such functions are 'facility-based', subdivided by 'management process'. Where projects are document-based, similar modules as above are required. For processes/ projects, a clear and holistic understanding of their nature is needed. The management processes in these are:

- to categorize, list and describe a facility;
- to analyse, model and evaluate the performance of a facility against indicators; and
- to appraise the facility against other options and make decisions about it.

A FIMS framework for these categories should comprise six main modules, namely:

- facilities (estate/assets) descriptive;
- facilities performance analysis and modelling;
- facilities option appraisal and decision-support;
- facilities report/document preparation;
- facilities report/document presentation;
- facilities report/document management.

FIMS FRAMEWORK IMPLEMENTATION

An analysis of existing organizational categories of computer-based infor-mation (as described in *Facilities Management '93*, pp. 150–5, referred to above) shows that three main component categories can be identified:

- drawing-based information;
- asset/estate-based information;
- services/process-based information.

The FIMS framework suggested above shows that these three categories fundamentally fit the 'facilities description module'. Data held in these categories mainly provides information about estate and assets in both graphical and textual forms. Often these three categories do not 'talk' with each other, as they are not integrated, and consequently gaps in information handling and management processes make decision-making more difficult. Furthermore, this data is not suitably structured to allow either data extraction or integration. The facilities description module should be geared to do this and to enable links to be made to the other two categories.

For each of the above modules – the constituent parts of the framework – there is a set of computer tools. These are CAFM tools or systems. It is important when evaluating and selecting such systems that they fulfil exact needs, wants and priorities, and that extraneous and unusable features are ignored. Any CAFM system should be assessed in terms of its suitability and efficiency in aiding the decision-making processes at strategic, tactical and/or operational levels. Significant technical criteria include:

- fast processing of complex data/information with maximum accuracy, and reliable maintenance and upgrading;
- flexibility with 'open structure' to allow for greater integration and exchange of data;
- comprehensiveness in terms of functions and features – the more applications that can be handled efficiently by one system the more appropriate it will be;
- speedy and user-friendly interface to ensure that the system is used to the full.

Most software is likely to meet core CAFM applications needs. Any one system that meets all of these needs should be looked at enthusiastically but with care.

ORGANIZATIONAL CRITERIA FOR A CAFM

CAFM systems which are suitable for certain FM application areas may not fulfil the exact requirements of an organization. The criteria adopted should not be supply led but demand driven and, as explained in *Facilities Management '93*, 'a system should be selected on the basis of how it fulfils certain FM requirements rather than on what the system offers generally.'

Crucial to FIMS operations is the identification of requirements for the availability and level of skilled personnel to manage and use such systems, e.g. skills in direct systems operation; in integrating and customizing systems for their full and more efficient use; in developing the component parts of an FIMS framework tailored to FM applications; and in systems management generally.

Inventories for comprehensive and up-to-date listing of facilities do not exist in many organizations. Such inventories covering all existing facilities being managed should form the basis for building up a FIMS. The implementation of effective FIMS procedures allows for more appropriate decision-making.

A lack of appropriate systems and application 'conventions' creates incompatibility between systems and hinders communication between departments. Standard common understanding, terminology and conventions (e.g. for layering) are essential for effective operations and management of FIMS. Finally, there is a lack of efficient exchange and integration of information, which is needed within and between different organizations, in electronic form. Such practices are caused by incompatibility of hardware, software, formats and data structures. This can lead to costly and confusing duplication of databases.

SOME CONCERNS

Finally, this chapter gives a brief account of some issues of concern in the field of information systems for facilities. The subject grows almost exponentially but, despite this, organizations wishing to obtain authoritative independent information of particular relevance to the automation of their FM functions, and the use of CAFM systems, find it hard to come by. Publications in the field of collective experiences of applied CAFM are sought after but particularly scarce. Published research findings in this area are rare too, though some are to be found in G.P. Keeffe's report *Corporate Computer-Aided Facilities Management: A Case Study*, published by Salford University in 1992.

Key concerns can also be identified regarding the future of CAFM in its main role of aiding the processes of strategic facilities management. There are two main sets of issues concerning the CAFM practice/market, namely 'demand' and 'supply'.

Demand-side factors

At a strategic level, there is a lack of recognition and understanding of the significance to business and operations of appropriate FM and FIMS practice. As Keith Alexander wrote in 1992 in his article entitled 'Quality managed facility' (*Facilities*, Vol. 10, No. 2, pp. 19–23): 'Properly managed FM information systems should be considered as organizational assets and resources rather than liabilities.' It is too easy to regard them as being 'peripheral' to 'core-business' and either be unaware of, or misunderstand, the cost-benefit analysis formula underlying their adoption.

Organizations are hampered by a lack of knowledge in two areas – how to effectively manage facilities operations, and how to structure related data/information for decision-making processes. Together these can lead to an inability to identify and define organizational requirements for FIMS. The absence of feedback from organizations to suppliers causes many systems vendors to develop *ad hoc* systems based on their own interpretations and motives! P.J. Montenero wrote of this in his article 'Beware of the CON in some consultants' in the IFMA *Facility Management Journal* (January/February, 1991).

Supply-side factors

CAFM systems tend to be commercially oriented in a supply driven market with proprietary 'application systems' finding a niche in the FM field. With some customization and development work, some can manage to produce limited FM application modules – space or energy management systems etc. – and claim to be sub or full systems of CAFM. Ignorant or ill-prepared facilities managers may well be duped. Users need to take the lead in defining CAFM systems development, as both Joudah and Montenero have pointed out above.

In summary, on the supply side of FIMS, most developers of software packages do not consider the strategic requirements of FM organizations. On the demand side, concerns focus on the strategic level of FM. Appropriate and timely education and training can improve this situation for system vendors and developers, and executives and other key decision-makers in user-organizations.

CONCLUSIONS

From the above, it can be clearly seen that the role of information management in the operation, or practice, of facilities management at all levels is of crucial importance. Research and benchmarking activities can establish a comprehensive set of criteria for appraisal that is truly 'responsive' to the varying and changing needs and requirements of FM practice. Findings from such activities can be of benefit both to organizations in helping them to define their exact requirements, and to systems vendors in defining a way forward for appropriate systems development.

Given the continuous technological advancement and complexity of IT technology and products, organizations (both users and vendors) could suffer from *ad hoc* FIMS policies and the absence of appropriate FIMS strategies. Premature obsolescence and the incompatibility of systems can result in very costly and inefficient operations.

SUMMARY

Business can be characterized by three predominant concerns – improving customer orientation, supporting creative workers and realizing the potential of information technology. Its complexity increasingly relies on the skills and knowledge of the right people. Without them, companies cannot expect to deliver the levels of competence, problem-solving and courtesy demanded by customers.

A computer-aided facilities management approach to the strategic requirements of an organization establishes the criteria for the selection of appropriate systems, and for incorporation with facilities information management systems to provide management information for strategic decision-making. Facilities organizations which aggressively utilize information technology not only transcend the price competition, which ensnares most of the profession, but also provide services which have more value to their end user clients as well. The rising expectations of these users – the customers – mean that they are better informed, have more choice and therefore have more power.

Developments in the field of computer-aided facilities management (CAFM) may be hampered by not considering the strategic requirements closely enough. Most organizations want the benefits of information technology, but not all are willing to allocate top management time and attention to achieve it. Those which do are not surprised to see their businesses and professional lives transformed.

BIBLIOGRAPHY

Essential reading

Hammer, J.M. (1988) *Facility Management Systems*, Van Nostrand Reinhold, New York.

Negreponte, N. (1995) *Being Digital*, Hodder & Stoughton, London.

Teicholtz, E. (1992) *Computer-Aided Facility Management*, McGraw-Hill, New York.

Wysocki, R. and Young, J. (eds) (1990) *Information Systems: Management Principles in Action*, John Wiley & Sons, New York.

Recommended reading

Clutterbuck, D. (ed.) (1989) *Information 2000: Insights into the Coming Decades in Information Technology*, Pitman, London.

Gerstein, M.S. (1987) *The Technology Connection: Strategy and Change in the Information Age*, Addison-Wesley, Reading, MA.

Hamer, J. (1991) 'Getting a jump on CAFM', *Facility Management Journal*, International Facility Management Association, January/February.

Keeffe, G.P. (1992) *Corporate Computer-Aided Facilities Management: A Case Study*, Salford University.

Stout, T. (1990) *The Financial Impact of Computer-Assisted Facilities Management Systems*, FM International Conference Proceedings, Centre for Facilities Management, University of Strathclyde.

Zuboff, S. (1988) *In the Age of the Smart Machine: The Future of Work and Power*, Basic Books, New York.

Further reading

Allen, T. (1977) *Managing the flow of technology*, MIT Press, Cambridge, MA.

Atkin, B. (1988) *Intelligent Buildings: Applications of IT and Building Automation to High Technology Construction Projects*, Kogan Page, London.

Blackler, F. and Oborne, D. (1987) *Information Technology and People: Designing for the Future*, The British Psychological Society, London.

Butler Cox (1989) *Information Technology and Buildings: A Practical Guide for Designers*, Butler Cox plc, London.

Editorial (1987) *Intelligent buildings*, Unicom Seminars Ltd, Conference Proceedings, Marble Arch, London.

Editorial (1990) *Facilities Management Tools*, Facilities Management International Conference Proceedings, Centre for Facilities Management, University of Strathclyde.

Information Technology Economic Development Committee (ITEDC) (1987) *IT future. IT Can Work: An Optimistic View of the Long-Term Potential of Information Technology for Britain*, HMSO Books, London.

Montenero, P.J. (1991) 'Beware of the CON in some consultants', *Facility Management Journal*, International Facility Management Association, January/February.

Peverley, D. (1992) *Management Responsibilities in Successful Automation Implementation*, reference paper by Integration and Product Services, Autotrol, Denver, USA.

Schrage, M. (1990) *Shared Minds: The New Technologies of Collaboration*, Random House, New York.

Vallely, P. (1994) 'Cure for an infomaniac', *The Daily Telegraph*, 4 January.

11	Support services

OVERVIEW

Customers will pay a lot for top quality. Firms that provide that quality will thrive – workers in all parts of the organization will become energized by the opportunity to provide a top quality product or service.   Tom Peters

The overall strategy for delivering services to customers can be described by the term 'service vision'. When organizations operate in a service-oriented economy, service vision is an important requirement for managers. It embraces marketing services, developing a unique way of providing the service (service concept), working to an operating strategy and having a service delivery system. Organizations need to recognize that their employees are their most valuable asset in their fight for advantage in any highly competitive marketplace. They may be the first point of contact that a customer has with an organization's core product or service. It is vital, therefore, that this contact is optimized by removing any distractions from the employee serving that customer.

Support services are conventionally categorized as overheads, but can be given priority according to the role they play in support of key business activities. For example, direct services support the front-line activities, whereas administrative overheads support internal customers. They can also be rationalized as essential and non-essential services. Basic services are delivered to service levels agreed with customers – clients and representatives of end users. Their delivery should add value to the services being offered.

Planning for support services must start by identifying the services needed to support the enterprise/operation. In addition, the user's requirements and the dimensions of service quality for a customer-focused internal organization, together with the techni-

ques for negotiating, evaluating and monitoring service performance, all need to be identified. In commerce or industry, general
services support the production and delivery of a product or
service. In the public sector they are delivered either direct to the
community or to support direct services and the organization itself.

Developing customer orientation in facilities management can be
an extremely complex process, as there may be a multitude of
internal and external customers. The juxtaposition of the needs of
those working in the building and external customers using the
building, together with statutory requirements and wider social and
environmental issues, can create conflicting needs which have to be
resolved.

A building performance-type approach can be used to derive
criteria against which the service may be appraised. This entails
determining user requirements and identifying the needs that can
be met from current resources. Service and functional requirements
are derived from this base and the performance specified. It is
against this that potential providers can offer to deliver services
and justify their performance. Service appraisal will include evaluation by the provider and surveys of user response and satisfaction.

Service quality is determined by the capability, approach and
credibility of the provider. These are the main components of
service quality and need to be identified and managed.

Keynote paper

Facilities management support services
Alan Kennedy

INTRODUCTION

Providing support services is a complex and evolving business – the
more so in a marketplace that is constantly changing. The term 'support
services within facilities management' needs to be clearly defined, particularly when the service activities themselves are capable of being
widely interpreted. Support services are of many kinds. One large UK
company identified some 58 skills among its in-house staff who were
providing support services, and this was by no means a comprehensive
listing, since it did not include communications or records management.

Support services may be provided by in-house resources, by external
contractors, or by a combination of the two. What they consist of and

how they are provided is often based on an organization's historical growth pattern. For new companies, or those occupying new buildings, support services are usually, though not always, planned around the building's design and operational concept.

Older organizations, with decades of company in-fighting and personal empire growth behind them, will often display peculiar and seemingly illogical responsibility groupings of support services. None of them, however, can escape the need to power, heat, light, clean, maintain and service properly the individual needs of the property's occupants, however they may group their support services to achieve this end. We can therefore seek to distil from the myriad of skills and formats associated with the term 'support services' certain core activities.

CORE ACTIVITIES

There are certain fundamental activities common to most facility managed organizations. These are:

- accommodation planning;
- administration;
- cleaning;
- communications;
- health and safety;
- property maintenance;
- records management;
- security.

Of course this is not all, but those listed above, taken in their broader sense, are usually found in most organizations, large or small. Other facilities management support services which may be a little less common include:

- car fleet control;
- catering;
- contract control;
- external property administration;
- internal/external landscaping;
- porterage;
- purchasing;
- reception;
- reprographics and stationery.

No doubt there are others. It all depends, as was noted earlier, on how the organization has evolved in its allocation of managerial responsibilities, and whether this was done in a strategically planned manner or as simply a historical legacy.

SUBSETS OF SERVICES

Within almost every one of the main support service headings above we can find subsets. As an example, the subset for cleaning might be:

- kitchen cleaning;
- office cleaning;
- special deep cleaning for carpets;
- waste disposal;
- window cleaning (external);
- window cleaning (internal);

and so on. It is not unusual to find organizations which manage individual contracts for each of the subset activities, although most are grouped as specialist support services under a lead contractor – in effect under one heading.

Where financial pressures on organizations have led to restructuring, a lot of attention is paid to reductions in overhead costs. Support services are usually prime targets for examination in some way – usually financial. Sometimes false economies are practised, but usually the aim is to amalgamate services and achieve cost reduction by outsourcing to companies specializing in providing both expertise and efficiency. Or at least that is the claim!

HOW CAN SUPPORT SERVICES BE PROVIDED?

Support services can, and should, be provided in accord with the two major drivers concerned, namely total quality management (TQM) and incentive-based performance standards. It can be contended that there are only three basic ways to provide FM support services:

- Do it with in-house staff.
- Contract it all out – sometimes known as total facilities management. Or:
- A combination of the two.

Of course there are variations of the above. Most organizations experience variations on how they operate these services and much depends on individual circumstances and market trends. The traditional way used to be to employ staff in-house to carry out support services. About ten years ago pressures grew on direct employment practices and they began to be eroded, slowly at first, by new external service providers. These were the first of the specialist FM companies and some operated on a national basis, such as Forte in catering. Business amalgamations, improvements in national communications and a greater willingness of service staff to relocate within the UK to further their careers have led to a quickening of the pace of support service contractor growth.

Some years ago, the first real sign that facilities management was a recognized support service industry became indisputable. The specialist associations and institutions began to raise the profile of the industry. This greater attention assisted the provision of support services to a much wider market. Facilities management support services are respected, in demand and often seen as adding value. This will be so as long as those who practise FM provide value for money and demonstrate improvements in the way support services are carried out. But with every market opportunity come the charlatans. Beware of those who hawk shallow expertise and cut-price labour resources. As ever, you pays your money and...!

ADDRESSING THE ISSUES

The range of issues being addressed in facilities management are changing and expanding. Whereas previously they were all about energy efficiency, space standards, workstation design, lighting densities, etc., they have been joined by issues of partnering with contractors, service response systems, performance standards, value-added gains and win-win situations.

The provision of the support services marketplace therefore continues to grow and become broader. One reason for this is the increased complexity of relationships between the purchasers (buyers) and the providers (suppliers) as they evolve to meet the demands of dynamic businesses. A number of key issues therefore need to be identified and addressed during any decision-making process for the provision of support services.

Quality of service

The drive for service comes from within a client organization as it inevitably becomes involved in developing quality systems for the delivery of its business to its customers. At the same time the organization will find there is an increased range of contracted services on offer in the market-place. Not least is this evident from some work done by the Centre for Facilities Management at Strathclyde University and published in November 1992 as *An Overview of the Facilities Management Industry – A Statement*. The delivery of support services to the organization cannot be excluded from an organization's quality journey.

However, traditional adversarial relationships are not conducive to reducing quality costs. Nor do they strive for continuous improvement, or seek solutions rather than apportioning blame. Completely different relationships have to be developed between purchasers and providers, or buyers and suppliers as they are sometimes called, regardless of whether it is an internal transaction between departments or an external contract.

The incentives for developing relationships such as partnerships are based on the dimensions of a quality service. According to L.L. Berry *et al.* in 'A conceptual model of service quality and its implications for future research' (in *Managing Services Marketing*, London Business School, 1989), these include reliability, understanding of the customer, security and communication between parties. The provision of support services has to be based on the worth that the service brings to the organization in terms of increased satisfaction, productivity and motivation. In other words, they should be services that add value and give value for money.

Support systems

What is meant by a support system as opposed to the core activity of an organization? Although it can be argued for strategic reasons that support and core are one and the same, for operational purposes the classification of service between support and core is best confined within one organization. What is peripheral to one will be central to another. For example, catering in a hospital will be far closer to health care than in an office where the activity is insurance broking.

A support service may be one which plays an underlying role in an organization rather than centre stage, but like all good productions the leading players cannot perform until the rest of the cast are also playing their parts, however minor. Individual support systems do not exist in isolation within an organization however they are provided. They overlap in terms of space allocation (e.g. storage), timing (e.g. when activities can be carried out to minimize disruption) and customers (many services will serve the same client base). Interactions between services should therefore be coordinated to meet users' needs.

Defining user requirements

The interpretation of the expectations of the customer is at the heart of FM. It requires expertise and is not a job for the faint-hearted. If beauty is in the eye of the beholder and beauty is a requirement of the user, then it can be a very difficult job! It is especially so when a customer requires a level of service which will take twice as long to provide as the provider had anticipated.

The liaison between the client and the provider is never straightforward, but it is crucial for the satisfactory delivery of a service. However, even 'satisfactory' may not be a good enough criterion. It depends on the organization's definition of quality – whether it is fitness for purpose (utility), or to be the best (excellence), or in TQM terms both of these.

There may be gaps in the perception about service quality. This can be at an operational level between the end user and the delivered service or at a strategic level between the purchaser (the end user's representative) and the provider. Perceptual mapping techniques, such as that provided by the Pacesetter Group's 'Quality Map' software (written and published by the Group in Birmingham in 1990) provide a means of identifying where gaps in perception exist and how they can be bridged. An obvious constraint on this process is one of 'limited resources' – what the customer wants and what he is prepared to pay for may produce another chasm for service quality to fall into if there is no common understanding of what is achievable.

Performance and service level agreement

The interpretation of the user's requirements is formalized into agreements of the standard of service required. These standards may be prescriptive (i.e. a technical specification) or may be performance based. Such descriptions of a service are not mutually exclusive. Both a performance specification and a technical specification may be required. The performance specification may set the standard required while the technical specification provides the detail. For example, the performance standard for catering may be to serve 500 people lunch between 12.00 and 14.00 and provide three choices of starter, main course and dessert. The technical specification may contain the recipes for each choice on the menu.

The construction of a service level agreement is required between the parties to determine the standard of service required and how conformance will be measured. It provides the mechanism for determination, and for both parties to the agreement to be audited to ensure satisfactory delivery and receipt of services.

MEANS OF PROVIDING SERVICES

Agreements between parties may depend on the relationships which are developed between purchaser and providers of services and the degree to which the service relies on hardware (i.e. control systems and machines) or software (people, management skills). Below are six examples of service groups.

Hospitality services

The services defined here are those confined to reception services: they do not include entertainment services. The differentiation in this type of service is between the hardware – security control systems, the space, etc. – and the software – people and procedures, etc. The quality of service will depend on its purpose and the atmosphere that is to be created. Many organizations require tight control over the use of buildings – government buildings and secure installations are two examples. The quality of such services may then be measured in terms of the numbers of people both logged in and out of the system during a visit.

The hardware and software in hospitality services often combine to create an impression to users. The reception area may have the most elaborate finishes in the whole building to provide the right ambience and convey the values of the organization to visitors. However, if the requirement of the reception is to welcome people the environment alone may not be sufficient. The personality of the receptionist then may be a determining factor. This is an important consideration where, to reduce costs, people are increasingly being replaced by machines.

Property services

Property services include ongoing maintenance, replacement of components of the building, estates services (rents, rates, etc.), accommodation planning and special 'projects' such as refurbishment, relocations or extensions. There is often an in-house department which plans these activities but they may subcontract areas of work – specialist maintenance, for example. Sometimes they call in consultants for specific tasks which occur periodically and require expertise outside the experience of in-house staff – space planning, disposals and acquisitions are cases in point.

Administrative services

The provision of such diverse services as a 'help' desk, mail and courier services, record management, general housekeeping (including cleaning and landscaping) and so on fall into this category. These may be contracted out as a bundle of services which are billed back to the client and based on demand. However, there is a trend towards service level agreements for this type of service so as to cut down the administrative burden for the client organization. Such agreements do give problems for a contractor because demand is unpredictable and pricing has to be carefully structured.

Contracts for administrative service provision can be competitively tendered or they may be negotiated on the level of service required. The criteria for negotiation can include response times, volume, choice of stock for stationery, hours of operation and so on. Negotiation may sometimes replace competitive bidding for contracts on the basis of these criteria, but organizations may prefer to choose suppliers on other criteria such as providing sources of income to the local economy and nationwide deals for all branches of the organization (single source suppliers).

Amenity services

These include such services as catering and vending, childcare and fleet management. The style of provision will depend on the type of organization and the expectations of customers. Catering facilities, for example, can range from silver service restaurant facilities to self-service canteens, but their use and pricing structure will depend on their perception in terms of the organization's objectives. Many companies subsidize meals for staff or provide them free of charge because they feel that it is a prerequisite for attracting high calibre staff when skilled personnel are in short supply. Also there may be several locations within an organization depending on its structure – there may be separate catering facilities for executives, for example.

Vending machines have replaced tea ladies and their trolleys, with the aim not only of reducing costs but of providing round-the-clock service for organizations which operate outside normal working hours. Sometimes these services are provided as tax benefits to employees, as is also the case with the provision of a company car. Another trend is to provide creches for the children of staff as part of the salary package. The Broadgate Development in London is an example of this.

Environmental issues are receiving more attention from organizations wishing to portray a green image. Office landscaping, recycling schemes and designated smoking areas may be token gestures by companies towards their employees. Broader issues range from the development of environmental control systems which enable individuals to change the condition of their working environment, to the elimination of CFCs from air-conditioning equipment.

Communication services

Communication services are growing in complexity. Networks of communication are set up internally by e-mail services on computer, and by phone. External connections are made through modems, fax and telephone systems. These networks often require specialists to install and maintain them. Flexibility is being built into systems. PABX facilities can be introduced: with under-floor links able to give individual users connections to several types of communication at a single point, these are becoming more common. In such cases, floor boxes which can be moved to suit different layouts are provided. Purchasers of these services are looking for reliability (of data and equipment), minimum maintenance and no disruption to the business.

Total facilities management services

'Total facilities management' can be defined as a single complete package of services provided by one organization to another. It has the advantage of achieving one interface between purchaser and provider, and it frees the client organization as there is only one bill to pay. The total FM route is being sold to organizations wishing to reduce staff and costs. It can provide an opportunity for a close relationship to develop between the parties.

The disadvantages of a total FM package are perceived loss of control by the client, the providers not understanding and accepting the organization's culture, and a mismatch between the perception of the purchaser and what is delivered by the provider. TFM may only give a different management style with different costs as staff previously employed in-house may be re-employed with new terms and conditions by the contracted organization.

CONCLUSIONS

The provision of support services in an organization should not be determined by what suppliers are willing to provide, but by the needs of individual organizations. It should be demand led, with needs based on the organization's business, as well as its goals and style of management. Relationships, whether contractual (a total FM contract, partnering, single or multiple sourcing) or within the organization, should be developed as a way of allowing both providers and purchasers to strive towards the same quality gaol – continuous improvement of the service to meet the needs of the business.

SUMMARY

A support service approach to the provision of facilities requires continuous adaptation to changing customer requirements. Managing customers entails working with their perceptions and expectations, so it is essential to conduct regular surveys of their attitudes and to listen generally to them. It is from customer perceptions of service quality that measurement systems can be derived and because of this, service providers should always endeavour to delight customers by exceeding their expectations.

If organizations recognize how valuable their staff are in a competitive world, and appreciate their role in fronting their products or services, they can more readily optimize the situation by removing any distractions from the employee serving that customer.

The promotion of workplace health has to be designed into the customer service process as a vital strand of the corporate business strategy. Poor working conditions, or a perceived uncaring employer, may affect the way in which a customer is treated which will eventually affect 'sales'. Employees should also be thought of as customers of the organization who can expect to be provided with a hazard-free working environment as the starting point – and the best setting – for their work processes in support of customers.

Providing an effective level of customer service and support is a major business challenge. If undertaken successfully, it can help ensure customer loyalty, win new business and encourage effective competition.

BIBLIOGRAPHY

Essential reading

Armistead, C.G. and Clark, G. (1992) *Customer Service and Support: Implementing Effective Strategies*, Financial Times/Pitman, London.
Carlzon, J. (1987) *Moments of Truth*, Ballinger, New York.
Friday, S. and Cotts, D.G. (1995) *Quality Facility Management*, Wiley, New York.
Zeithaml, V.A., Parasuraman, A. and Berry, L.L. (1990) *Delivering Quality Service*, Free Press, New York.

Recommended reading

Burnett, K. (1992) *Strategic customer alliances*, Pitman, London.
Gerson, R. (1993) *Measuring Customer Satisfaction*, Kogan Page, New York.
Hollins, G. and Hollins, B. (1992) *Total Design: Managing the Design Process in the Service Sector*, Pitman, London.
Horovitz, J. and Panak, M.J. (1992) *Total Customer Satisfaction*, Pitman, London.

Schonberger, R. (1990) *Building a Chain of Customers*, Hutchinson, London.

Zemke, R. and Schaaf, D. (1989) *The Service Edge*, New American Library, New York.

Further reading

Blanding, W. (1985) *Practical Handbook of Customer Service Operations*, Transport Press, London.

Denton, K.D. (1991) *Horizontal Management: Beyond Total Customer Satisfaction*, Lexington Books, New York.

Donovan, P. and Samler, T. (1995) *Delighting Customers*, Chapman & Hall, London.

Hiles, A.N. (1993) *Service Level Agreements*, Chapman & Hall, London.

Hinton, T.D. (1991) *The Spirit of Service: How to Create a Customer-Focused Service Culture*, Kendall/Hunt, Dubuque, Iowa.

Lash, L. (1989) *The Complete Guide to Customer Service*, Wiley, New York.

McNealy, R.M. (1995) *Making Customer Satisfaction Happen*, Chapman & Hall, London.

Peters, T. (1989) *The Customer Revolution*, The Economist Conference Unit, London.

Taylor, L. (1992) *Quality: Total Customer Service*, Century Business Books, London.

Thompson, W.T., Berry, L.C. and Davidson, P.H. (1978) 'Managing markets through planning', in *Banking Tomorrow*, Van Nostrand Reinhold, New York.

<table>
<tr><td>

12

</td><td>

Project management

</td></tr>
</table>

OVERVIEW

The overall planning, control and coordination of a project from inception to completion aimed at meeting a client's requirements and ensuring completion on time, within cost and to the required quality standards.   Chartered Institute of Building

Facilities management is the continual process of planning, providing and managing facilities through the changing needs of an organization. There will be occasions within this process when a particular activity can be defined as a project and project management techniques can be applied. To some the distinction between facilities and project management is ambiguous. However, a project has a defined start and end, and is controlled, usually, through a contract with finite resources committed for its execution.

Projects, of varying size, value and complexity, are a common feature in the facilities management field and need specific management skills to handle them. These skills relate to techniques which have to be applied to an ever changing and competitive environment as projects are initiated, developed and completed. Organizations, in business, industry and government, recognize the benefits of management by project, not least in order to minimize the business disruption associated with, what are often, major pieces of work.

Facilities management, therefore, entails responsibility both for projects and for the other normally continuing operations. The facilities team, at the operational level, must continue to fulfil its own specific responsibilities for planning and delivering services to support business effectiveness. It is the projects which have to be managed whilst services continue and disruption to occupancy and the effect on production is minimized. The facilities team together with any separately identifiable project management teams have to work in tandem, an arrangement which requires continuous communication.

Construction and installation projects, as well as major refurbishments and relocations, are typical of the major corporate projects normally led by the facilities management team. Such projects combine logistic and resource management, with complex social factors associated with using plant and buildings, as well as moving personnel and even, on occasion, their families. The facilities team continues to hold prime responsibility for all 'customer care' throughout such projects and therefore requires a mix of technical and human resource management skills.

During major projects, an organization is effectively changed from one state to another and this process of change needs to be managed. Sometime those responsible for bringing about change are appointed for the duration of the project and the team responsible for any parallel service delivery phased in to take over on completion. The discipline of 'interim management' has emerged to focus on and improve the processes of change.

The demands of simultaneously managing a project, ensuring minimum disruption to operations and setting up a project team require advanced organizational skills so as to minimize the impact on the organization. Respective responsibilities must be clearly defined and resources allocated to ensure overall success. It is essential that the project has commitment from a senior manager with the authority to commit the necessary resources. A project sponsor needs to be appointed to represent the key interests of the client organization and provide a timely response for decisions.

Keynote paper

Project management

Stewart Wood

INTRODUCTION

Facilities management as a function within any organization is a complex and wide-ranging responsibility. The differing activities that require attention through the normal working week vary enormously from the strategic to the operational, and from the mundane to the technical, as at all times the needs of the core business require to be supported effectively. Given this, the addition of a 'project' into this environment is not to be taken lightly.

Projects come in all shapes and sizes, and vary in their complexity, cost and duration. It is important to thoroughly analyse their true depth and scope to ensure that the correct level of expertise is applied. 'Success' for most projects is easily defined by the brief statement, 'on time – in budget'. This disguises much frustration, negotiation, re-evaluation, communication, hard work, etc. that, almost by definition, will be involved. All project activity is normally condensed into a set period of highly polarized activity which dominates daily life and demands constant attention.

If it is allowed to, project management has the unique habit of absorbing efforts to the detriment of everything else, so even greater care must be taken in the approach to project planning to ensure this domination does not affect the smooth running of the rest of the business. The importance of the project to the business will dictate its priority, and how ready the business will be in accepting the disruption it will bring. The staff of the business, referred to in the rest of this chapter as customers, will tolerate more providing the communication with them about the project is well structured and they are able to appreciate benefit in the long run to the business or to their own environment. It is seldom that the luxury of a project in a 'greenfield' environment is possible.

Together with the importance of customers comes that of business continuity and the overwhelming need for the safe access to and egress from the project site to be maintained at all times. To this end a thorough planning exercise must be carried out at a very early stage, to ensure programming takes full account of all known activities for the duration of the works. These activities can take many forms, from known special events, through to high workload peaks and important dates in the calendar for the department or company. As part of this, activities associated with any other parallel projects must be avoided.

CUSTOMER SATISFACTION

As project success can be measured in many ways, it is best to agree how customers will determine the success of a project and strive to document what the desired outputs will be. There should be a review to set the objectives of the work and this then needs to be embodied in the project plan. This is to show what critical success factors can be looked for as a result of the efforts. Setting expectations at the outset is important, as it is too easy to lose track of what is to be achieved when business pressures can change requirements constantly. The components of success are important for all concerned – this is not to narrow the requirements, but to ensure adequate recognition of major change. Good management of this element will ensure the levels of expectation are achieved and recognition duly forthcoming for a job well done. Trying to achieve this after the event is impossible. The main drivers of a project will always be cost and programme, but a project manager has also to ensure the satisfaction of the customer.

COMMUNICATION

The need for good effective communications throughout a project is important to the success of the works. It is an essential if the project drivers are to work. It is necessary to determine firstly who needs to know, who would be best to know, and who will assist in the delivery of the work by being kept informed. Finally, there is the need for everyone's general interest to be satisfied through regular briefing specifically designed to inform. The communication plan should form an early part of any planning exercise and it should be tested at regular intervals. Good, timely communication can smooth out inconveniences as they tend not to be readily accepted unless understood and the benefits explained. Being seen to be approachable through regular communication can stop time being wasted. Handled effectively, this will not be seen as weakness but can mean customers will feel involved. Early information can also help prevent mistakes as a wider audience has knowledge of the activity. It may also ultimately avoid downtime or costly rework.

AFFORDABILITY

Projects usually stem from an overwhelming need for some form of change required by the business, group, department or organization. Needs may come from product change, legislation change, staff need, relocation or a host of other reasons. Seldom is this a frivolous requirement, therefore value for money will always be the watchword.

Work on a project has always to be affordable and therefore communication assumes even greater importance. There will always be an owner or client who will have the authority to commit the funding necessary for the work, or the changes that may be necessary during the course of it. The speed and effectiveness of this line of control will also drastically affect the overall price of the finished product. Having established the requirements and set the customer expectation level then the real task of evaluating the out-turn cost against the requirement can begin. Inevitably this will require some adjustments to be made to create a fit. Adjustments should be kept to a minimum in any project, but the whole plan should take into account what other funds are available for related or adjacent works that can be included in the requirement to get that extra positive contribution. This includes reviewing whatever asset management funds and plans, major repair reserves or any other planner's funds there are to see what might be available to help get a much more effective job. Good communications can help secure finance director support and a good facilities/project manager can bring fresh ideas to bear to give extra added value. All this is required at this point prior to tendering for the works.

TENDERING

There are many views on tendering and as many forms of tender. They vary from the requirement of public sector bodies to follow the *European Journal* process, through to single sourcing via a known and trusted design and build route. Each company or department will almost certainly have a process that must be followed, though that should not preclude a thorough appraisal of each project to pick and propose the most appropriate route. Much will depend on the complexity and value of the works to be undertaken as well as their context. Tendering is a very important part of the project process and part of it involves the selection of the right consultants, the right contractors and the right contractual relationship commensurate with the project requirements. Access to quantity surveying or similar professional advice may prove an advantage in assisting and guiding this process prior to making commitments.

Small projects may well be handled internally through a direct works arrangement, whereas the more comprehensive projects will require a competitively tendered specification, but, in any case, the need is to be able to select the appropriate quality and cost to suit the project needs. It will depend on the particular tendering procedures used as to whether further post-tender negotiation is allowed. If it is, it should be used to ensure requirements are understood and all aspects of the job allowed for, and any further cost reductions realized. This arrangement amounts to negotiation, which is never easy and must always be used sensitively and sensibly. It should not be, but in reality it is often the one opportunity to make the finances and project fit each other! Working together as a team can help to achieve the best deal possible to deliver the project.

THE PROJECT

The next stage is the task of making the whole project run smoothly to deliver what it was set out to achieve. Given all the preparatory work the basic planning should manage this effectively all the way through to delivery of the final project. The satisfactory management of the delivery is a matter of constant vigilance starting from a site briefing, a full job-safety review – defining all the operating restrictions – and clarifying how works will be carried out. The need for a clear understanding of the safety procedures that will prevail is essential, as they will be enforced by the health and safety at work legislation.

To fully understand the contract, a list of questions should be addressed which can help set the scene. Such questions are as follows.

- What is legally required?
- Who will be in charge on site?
- How will the work be carried out on site?
- What are the site boundaries and means of access and egress?
- Will contractor staff have access to occupied areas?
- What safe working practices will prevail?
- What 'permits to work' will be required and who is the authorized person?
- Who will sign them, and who is his/her deputy?
- What electrical interruptions will be required and who will be affected?
- What lead times for outages are required?
- What progress meetings are required?
- Who will chair these sessions, minute them and follow up on outstanding actions?
- Has a risk assessment been carried out and areas of concern identified?
- How will a clear 'kick-off' be managed to ensure proper induction to the site?
- How will security be maintained?
- Is there a 'housekeeping' policy and who will enforce it?
- What certificates will be required on completion?
- How will handover be managed?
- What commissioning will be carried out?
- What will be the means of payment – stage, progress or other?
- What percentage retentions will be held?
- What information is wanted to subsequently manage the finished project?

This is by no means a comprehensive list, but serves to indicate that there is a real need to plan, prepare, predict, publish and participate fully to make the project a success. Although the list is of what might be called the tangible elements, the intangible factors require as much attention. When contemplating an innovative approach or solution, even more attention to the delivery is required.

Once on site, the project assumes a new dimension as the fruits of careful planning start to materialize. The temptation is to over-manage the activity and thus get in the way of the team who are there to deliver. This must be resisted at all costs to avoid legitimate blame for any delays that may materialize. Be aware that there will be changes of heart, of mind and of consequence along the way. A flexible approach can cope with this while maintaining direction. At an agreed point design changes will require to be frozen as further change will escalate cost, so a firm and robust attitude is needed. Having effectively planned the whole process and communicated it widely, changes should be minimal.

A critical path analysis should be part of the preparation, enabling regular review of progress to ensure the project is on target. Any slippage needs to be quickly dealt with to avoid disaster. The performance of the project team should be regularly evaluated to ensure good working relationships are being maintained. A well motivated and educated team can pay enormous dividends in dealing together with inevitable problems. An adversarial attitude in a team will mean problems escalate and progress is lost. More is achieved through healthy negotiation and professional harmony than is ever achieved through recrimination or litigation. The ability to walk a fine tightrope is a useful aspect of a facility/project manager's personality.

COMPLETION PHASE

Keeping on track and maintaining momentum is one thing, but handling the rigours of completion is another. This is the point where the greatest vigilance is required as everyone involved tries to complete their work. Early efforts at planning and communication come into play as expectations have been set. Certificates have to be received, CAD drawings must be presented, commissioning figures are required to be published and documents handed over, and finally design standards have to be witnessed. Nothing should be taken for granted. Everything needs to be checked and checked again, especially against items on earlier checklists. This avoids problems later on should any failure occur. Through progress meetings and the accurate record-keeping, minuting and publication of critical path based achievement a thorough understanding of the completed work is achieved.

With such a record system, coping with essential feedback is that much easier. It helps with the final account and minimizes the length of the snagging list. Assessment of customer satisfaction then follows in the form of a post-project review. This should be done for all projects to ensure lessons learned throughout the process are used to good effect. The final yardstick of meeting cost and programme can be assessed. The importance of the project to the business will dictate its priority, and how ready the business will be in accepting the disruption it will bring. Customers will tolerate more providing the communication paths are well structured and used so that they are able to appreciate benefit in the long run to the business or their own environment.

Setting expectations at the outset is important, as it is too easy to lose track of what was set out to be achieved when the pressures of business have a tendency to constantly change the requirements. The components of success are important for all concerned. This is not to narrow the requirements, but to ensure adequate recognition of major change. Good management of this element will ensure the levels of expectation are achieved and recognition duly forthcoming for a job well done. A concluding completion of the customer expectation form and a check that he/she is content with the outcome completes the production stage of the project. A final 'did the customer get what he wanted at the price that could be afforded?' will only be answered in the affirmative if the project performs to expectations throughout its expected life.

CONCLUSIONS

The phases of project management that should result in a smooth delivery of works on site to the satisfaction of customers have been highlighted. The need for careful planning, communication, management of activity and safe working practice has been outlined, and finally a professional and thorough handover process has been emphasized. It is these elements which can ensure the most effective, harmonious and incident-free activity giving value for money which will be a credit to the facilities manager.

SUMMARY

A well developed battery of project management techniques is available for handling all stages of project planning and programming, cost control and the management of contractors. These tools and techniques, when applied to well defined elements and phases, aim to ensure that projects are delivered to quality, at cost and on time. They call for careful planning, communications and management of activities, as well as for safe working practices and a professional and thorough handover process.

Progress and productivity measurement and optimization are essential for the effective management of time in a project and practical tools are available to measure these throughout the project duration. Project time is an expensive resource and it has to be assumed that any overrun will automatically lead to increased expenditure or loss.

The essential skills of managing quality, value and risk apply to project management, just as they are at the heart of facilities management. A manager of a project will take responsibility for teambuilding and setting up effective communications. Proven strategies need to be employed to optimize productivity and to manage the project budget and resource constraints. Responsibilities for these management skills, for health and safety standards and for controlling environmental impact must be clearly delegated in order to avoid poor performance through confusion and indecision.

The application of project management techniques does not always lead to improved project performance. On completion of the project and its acceptance by the client, a full project appraisal should be carried out and the project evaluated against the original objectives.

BIBLIOGRAPHY

Essential reading

Barkley, B. (1992) *Customer-Driven Project Management: New Paradigm in Managing Total Quality*, McGraw-Hill, New York.

Kharbanda, O.P. and Stallworthy, E.A. (1991) *Project Teams: The Human Factor*, NCC, Blackwell, London.

Morris, P.G. and Hough, G.H. (1987) *The Anatomy of Major Projects: A Study of the Reality of Project Management*, Wiley, Chichester.

Turner, R.J. (1993) *The Handbook of Project-Based Management*, McGraw-Hill, New York.

Recommended reading

Borjeson, L. (1976) *Management of Project Work*, The Swedish Agency for Administrative Development, Stockholm.

Chartered Institute of Building (1982) *Project Management in Building*, CIOB, Englemere.

Harrison, F.L. (1986) *Advanced Project Management*, Gower, Chichester.

Raftery, J. (1993) *Risk Analysis in Project Management*, E & FN Spon, London.

Reiss, G. (1991) *Project Management Demystified*, Chapman & Hall, London.

Walker, A. (1989) *Project Management in Construction*, Blackwell Scientific, Oxford.

Further reading

Anderson, E.S., Glude, K.V., Haug, T. and Trevor, J.R. (1987) *Goal Directed Project Management*, Kogan Page, London.
Cleland, D.I. and King, W.R. (1983) *System Analysis and Project Management*, McGraw-Hill, New York.
Cleland, D.I. and King, W.R. (eds) (1988) *The Project Management Handbook*, Van Nostrand Reinhold, New York.
Cooper, D.F. and Chapman, C.B. (1987) *Risk Analysis for Large Projects*, Wiley, Chichester.
Goodman, L.J. (1988) *Project Planning and Management*, Chapman & Hall, London.
Lock, D. (1992) *Project Management*, Gower, Aldershot.
Sanders, N. (1991) *Stop Wasting Time*, Prentice Hall, New York.
Spinner, M.P. (1992) *Elements of Project Management*, Prentice Hall, Englewood Cliffs, NJ.

Further reading

MANAGEMENT

Aguilar, F.J. (1994) *Managing Corporate Ethics*, Oxford University Press, Oxford.

Albrecht, C. and Zemke, R. (1990) *Service America: Doing Business in the New Economy*, Warner Books, New York.

Allen, T. (1977) *Managing the Flow of Technology*, MIT Press, Cambridge MA.

Argyris, C.S. and Schon, D.A. (1984) *Theory in Practice: Increasing Professional Effectiveness*, Jossey-Bass, New York.

Armistead, C.G. and Clark, G. (1992) *Customer Service and Support: Implementing Effective Strategies*, Financial Times/Pitman, London.

Bartlett, C.A. (1989) *Managing across borders*, Hutchinson Business Books, London.

Beer, S. (1972) *Brain of the Firm* Herder & Herder, New York.

Burnett, K. (1992) *Strategic customer alliances*, Pitman, London.

Buzan, Tony (1992) *Use Your Head*, BBC, London.

Campbell, A., Devine, M. and Young, D. (1990) *A Sense of Mission*, Economist Books, London.

Cannon, T. (1992) *Corporate Responsibility*, Financial Times/Pitman, London.

Carlzon, J. (1987) *Moments of Truth*, Ballinger, New York.

Crosby, P. (1984) *Quality without Tears*, McGraw-Hill, New York.

Cunningham, I. (1994) *The Wisdom of Strategic Learning*, McGraw-Hill, New York.

Davidow, W. (1994) *The Virtual Corporation*, Harper Business, New York.

Dickson, G. (1989) *Corporate Risk Management*, Witherby, London.

Drucker, P.F. (1968) *The Practice of Management*, Pan/Heinemann, London.

Drucker, P.F. (1980) *Managing in Turbulent Times*, Heinemann, London.

Drucker, P.F. (1988) 'The coming of the new organization', *Harvard Business Review*, January/February.

Drucker, P.F. (1989) *The New Realities*, Harper & Row, New York.

Drucker, P.F. (1992) *Managing for the Future*, Butterworth-Heinemann, London.

Drucker, P.F. (1993) *Post-capitalist Society*, Butterworth-Heinemann, Oxford.

Easterby-Smith, M., Thorpe, R. and Lowe, A. (1991) *Management Research: An Introduction*, Sage, London.

Galbraith, J. (1973) *Designing Complex Organizations*, Addison-Wesley, Reading, MA.

Garratt, R. (1991) *Learning to Lead*, Harper Collins, London.

Garratt, R. (1994) *The Learning Organisations*, Harper Collins, London.

Garvin, D. (1988) *Managing Quality: The Strategic and Competitive Edge*, Free Press, New York.

Gerson, R. (1993) *Measuring Customer Satisfaction*, Kogan Page, New York.

Hamel, G. and Prahalad, C.K. (1989) 'Strategic intent', *Harvard Business Review*, May/June.

Hammer, M. and Champy, J. (1993) *Re-engineering the Corporation: A Manifesto for Business Revolution*, Nicholas Brealey, London.

Hampden-Turner, C. (1990) *Corporate Culture*, Hutchinson, London.

Handy, C. (1976) *Understanding Organisations*, Penguin Education, Harmondsworth.

Handy, C. (1984) *The Future of Work: A Guide to a Changing Society*, Blackwell, Oxford.

Handy, C. (1989) *The Age of Unreason*, Business Books, London.

Handy, C. (1992) *Inside Organisations: 21 Ideas for Managers*, BBC Books, London.

Handy, C. (1994) *The Empty Raincoat*, Hutchinson, London.

Hayek, F.A. (1988) *The Fatal Conceit*, Routledge, London.

Heller, R. (1990) *Culture Shock: The Office Revolution*, Hodder & Stoughton, London.

Kanter, R.M. (1984) *The Change Masters: Innovation for Productivity in the American Corporation*, Allen & Unwin, New York.

Kanter, R.M. (1989) *When Giants Learn to Dance: Mastering the Challenges of Strategy, Management and Careers in the 1990s*, Simon & Schuster, New York.

Kaufman, H. (1985) *Time, Chance and Organizations: Natural Selection in a Perilous Environment*, Chatham House, Chatham, NJ.

Kotter, J. and Hesketh, J. (1992) *Corporate Culture and Performance*, Free Press/Macmillan, New York/London.

Lahtinen, T. (1994) *Health, Happiness and Success*, Janus Publishing, New York.

Landow, W. and Malone, M. (1993) *The Virtual Corporation*, Harper Business Press, New York.

Lickert, R. (1961) *New Patterns of Management*, McGraw-Hill, New York.

Mayo, E. (1949) *The Social Problems of an Industrial Civilization*, Routledge, London.

Mintzberg, H. (1979) *The Structuring of Organizations*, Prentice Hall, New York.

Mintzberg, H. (1989) *Mintzberg on Management*, Free Press, New York.

Mintzberg, H. (1994) *The Rise and Fall of Strategic Planning*, Prentice Hall, New York.

Morgan, G. (1986) *Images of Organisation*, Sage, Newbury Park, CA.

Mullins, J. (1993) *Management and Organisational behaviour*, 3rd edn, Pitman, London.

Munro-Faure, L. and Munro-Faure, M. (1992) *Implementing Total Quality Management*, Pitman, London.

Naisbit, J. (1982) *Megatrends*, Warner Books, New York.

Naisbit, J. (1994) *Global paradox*, Nicholas Brealey, London.

Naisbit, J. and Aburdene, P. (1990) *Megatrends 2000*, Morrow, New York.

Olins, W. (1989) *Corporate Identity: Making Business Strategy Visible through Design*, Thames & Hudson, London.

Ormerod, P. (1994) *The Death of Economics*, Faber & Faber, London.

Pascale, R. (1990) *Managing on the Edge*, Penguin, Harmondsworth.

Peters, L. and Hull, R. (1994) *The Peter Principle*, Souvenir Press, London.

Peters, T. (1987) *Thriving on Chaos*, Pan Books, London.

Peters, T. (1992) *Liberation Management: Necessary Disorganisation for the 90s*, Macmillan, London.

Peters, T. and Austin, N. (1985) *A Passion for Excellence*, Collins, London.

Peters, T. and Waterman, R.H. (1981) *In Search of Excellence: Lessons from America's Best Run Companies*, Harper & Row, New York.

Porter, L.W. and Roberts, K.H. (1977) *Communication in Organisations*, Penguin, London.

Porter, M.E. (1980) *Competitive Strategy*, Free Press, New York.

Porter, M.E. (1985) *Competitive Advantage: Creating and Sustaining Superior Performance*, Free Press, New York.

Rees, W.D. (1988) *The Skills of Management*, Routledge, London.

Reich, R.B. (1991) *The Work of Nations: Preparing Ourselves for 21st Century Capitalism*, Simon & Schuster, London.

Schon, D.A. (1971) *Beyond the Stable State: Public and Private Learning in a Changing Society*, Temple Smith, New York.

Schon, D.A. (1983) *The Reflective Practitioner: How Professionals Think in Action*, Basic Books, New York.

Schonberger, R. (1986) *World Class Manufacturing: The Lessons of Simplicity*, Free Press, New York.
Schonberger, R. (1990) *Building a Chain of Customers*, Hutchinson, London.
Schumacher, E.F. (1973) *Small Is Beautiful: A Study of Economics as if People Mattered*, Blond & Biggs, London.
Schumacher, E.F. (1979) *Good Work*, Cape, London.
Schwartz, P. (1991) *The Art of the Long View*, Doubleday, New York.
Semler, R. (1993) *Maverick*, Hutchinson, London.
Senge, P.M. (1992) *The Fifth Discipline: The Art and Practice of the Learning Organisation*, Century Business, London.
Tapscott, D. and Caston, A. (1993) *Paradigm Shift: The New Promise of Information Technology*, McGraw-Hill, New York.
Toffler, A. (1970) *Future Shock*, Pan Books, London.
Toffler, A. (1980) *The Third Wave*, Morrow, New York.
Toffler, A. (1984) *Previews and Premises*, Pan Books, London.
Toffler, A. (1990) *Power Shift*, Bantam Books, New York.
Waterman, R.H. (1994) *The Frontiers of Excellence: Learning from Companies that Put People First*, Nicholas Brealey, London.
Zeithaml, V.A., Parasuraman, A. and Berry, L.L. (1990) *Delivering Quality Service*, Free Press, New York.
Zemke, R. and Schaaf, D. (1989) *The Service Edge*, New American Library, New York.
Zuboff, S. (1988) *In the Age of the Smart Machine: The Future of Work and Power*, Basic Books, New York.

FACILITIES MANAGEMENT

Alexander, K. (ed.) (1993) *Facilities Management 1993*, Hastings Hilton, London.
Alexander, K. (ed.) (1994) *Facilities Management 1994*, Hastings Hilton, London.
Alexander, K. (ed.) (1995) *Facilities Management 1995*, Blenheim Business Publications, London.
Argyris, C.S. and Schon, D.A. (1984) *Theory in Practice: Increasing Professional Effectiveness*, Jossey-Bass, New York.
Armistead, C.G. and Clark, G. (1992) *Customer Service and Support: Implementing Effective Strategies*, Financial Times/Pitman Publishing, London.
Aronoff, S. and Kaplan, A. (1994) *Total Workplace Performance: Rethinking the Office Environment*, WDL Publications, Ottawa.
Avis, M., Gibson, V. and Watts, J. (1989) *Managing Operational Property Assets*, University of Reading.

Band, W.A. (1991) *Creating Value for Customers*, Wiley, New York.

Bannister, J.E. and Bawcutt, P.A. (1981) *Practical Risk Management*, Witherby, London.

Barkley, B. (1992) *Customer-Driven Project Management: New Paradigm in Managing Total Quality*, McGraw-Hill, New York.

Barrett, P. (1995) *Facilities Management: Towards Best Practice*, Blackwell, Oxford.

Bechtel, R., Marans, R. and Michelson, W. (1987) *Methods in Environmental and Behavioral Research*, Van Nostrand Reinhold, New York.

Becker, F. (1989) *Corporate Facilities Management: An Inside View for Designers and Managers*, McGraw-Hill, New York.

Becker, F. (1990) *The Total Workplace: Facilities Management and the Elastic Organization*, Van Nostrand Reinhold, New York.

Becker, F. and Steele, F.I. (1995) *Workplace by Design*, Jossey-Bass, New York.

Bennis, W.G. (1966) 'Changing organisations', *Journal of Applied Behavioural Science*, Vol. 2, No. 3.

Binder, S. (1989) *Corporate Facility Planning*, McGraw-Hill, New York.

Bone, C. (1993) *Value Analysis in the Public Sector*, Longman, London.

Brand, S. (1994) *How Buildings Learn: What Happens after They Are Built*, Viking, New York.

Butler Cox (1989) *Information Technology and Buildings: A Practical Guide for Designers*, Butler Cox, London.

Cannon, T. (1992) *Corporate Responsibility*, Financial Times/Pitman, London.

Carlzon, J. (1987) *Moments of Truth*, Ballinger, New York.

Carmichael, S. and Drummond, J. (1989) *Good Business: A Guide to Corporate Responsibility and Business Ethics*, Business Books, London.

Carson, R. (1962) *Silent Spring*, Penguin, Harmondsworth.

Clift, M.R. and Butler, A. (1995) *The Performance and Costs-in-use of Buildings: A New Approach*, BRE Report, Building Research Establishment, Garston.

Copper, D. and Palmer, J. (eds) (1990) *The Environment in Question*, Routledge, London.

Crosby, P. (1984) *Quality Without Tears*, McGraw-Hill, New York.

Davis, G., Becker, F., Duffy, F. and Sims, W. (1984) *ORBIT 2: Organizations, buildings and information technology*, The Harbinger Group, Newark, CT.

Duffy, F. (1989) *The Changing City*, Bulstrode Press, London.

Duffy, F. (1992) *The Changing Workplace*, Phaidon, London.

Duffy, F., Laing, A. and Crisp, V. (1993) *Responsible Workplace*, Butterworth Architecture, Oxford.

Evenden, R. and Anderson, G. (1992) *Management Skills: Making the Most of People*, Addison-Wesley, Reading, MA.

Facilities Management International Conference Proceedings (1990) Centre for Facilities Management, University of Strathclyde.

Flanagan, R. and Norman, G. (1995) *Risk Management and Construction*, Blackwell, Oxford.

Friday, S. and Cotts, D.G. (1995) *Quality Facility Management*, Wiley, New York.

Gann, D. (1992) *Intelligent Building Technologies: Japan and Singapore*, Science Policy Research Unit, University of Sussex.

Garvin, D. (1988) *Managing Quality: The Strategic and Competitive Edge*, Free Press, New York.

Goumain, P. (1989) *High Technology Workplaces: Integrating Technology, Management and Design for Productive Work Environments*, Van Nostrand Reinhold, New York.

Hammer, J.M. (1988) *Facility Management Systems*, Van Nostrand Reinhold, New York.

Hollins, G. and Hollins, B. (1991) *Total Design: Managing the Design Process in the Service Sector*, Pitman, London.

Joroff, M., Louargand, M., Lambert, S. and Becker, F. (1994) *Strategic Management of the Fifth Resource: Corporate Real Estate*, Industrial Development Research Foundation, GA.

Kernohan, D., Gray, J., Daish, J. with Joiner, D. (1992) *User Participation in Building Design and Management*, Butterworth, London.

Kharbanda, O.P. and Stallworthy, E.A. (1991) *Project Teams: The Human Factor*, NCC, Blackwell, London.

Markus, T. *et al.* (1972) *Building Performance*, Building Performance Research Unit, University of Strathclyde.

Morgan, G. (1986) *Images of Organization*, Sage, Newbury Park, CA.

Mullins, J. (1992) *Hospitality Management: A Human Resources Approach*, Pitman, London.

Munro-Faure, L. and Munro-Faure, M. (1992) *Implementing Total Quality Management*, Pitman, London.

Negreponte, N. (1995) *Being Digital*, Hodder & Stoughton, London.

Oxford Brookes University (1993) *Property Management Performance Monitoring*, School of Estate Management, Oxford Brookes University and Department of Land Management and Development, University of Reading, GTI, Oxford.

Powell, J.A., Cooper, I. and Lera, S. (1984) *Designing for Building Utilisation*, E & FN Spon, London.

Preiser, W.F.G. (1989) *Building Evaluation*, Van Nostrand Reinhold, New York.

Preiser, W.F.G. (ed.) (1993) *Professional Practice Facilities Programming*, Van Nostrand Reinhold, New York.

Preiser, W.F.G., Rabinowitz, H.Z. and White, E.T. (1989) *Post Occupancy Evaluation*, Van Nostrand Reinhold, New York.

Prieser, W.F.G., Vischer, J. and White, E.T. (1991) *Design Intervention: Toward a More Humane Architecture*, Van Nostrand Reinhold, New York.

Rondeau, E.P., Brown, R. and Lapides, P. (1995) *Facility Management*, Wiley, New York.

Ruck, N.C. (ed.) (1989) *Building Design and Human Performance*, Van Nostrand Reinhold, New York.

Schon, D.A. (1983) *The Reflective Practitioner: How Professionals Think in Action*, Basic Books, New York.

Senge, P.M. (1992) *The Fifth Discipline: The Art and Practice of the Learning Organisation*, Century Business, London.

Sommer, R. (1969) *Personal Space: The Behavioural Basis of Design*, Prentice Hall, Englewood Cliffs, NJ.

Spedding, A. (ed.) (1994) *The CIOB Handbook of Facilities Management*, Longman, Harlow.

Steele, F.I. (1973) *Physical Settings and Organizational Development*, Addison-Wesley, Reading, MA.

Steele, F.I. (1986) *Making and Managing High Quality Workplaces*, Teachers College Press, New York.

Sundstrom, E. (1986) *Work Places: The Psychology of the Physical Environment in Offices and Factories*, Cambridge University Press, Cambridge.

Teicholtz, E. (1992) *Computer-aided Facility Management* McGraw-Hill, New York.

Teicholtz, E. and Ikeda, T. (1995) *Facility Management Technology*, Wiley, New York.

Tompkins, J.A. (1984) *Facilities Planning*, McGraw-Hill, New York.

Turner, R.J. (1993) *The Handbook of Project-Based Management*, McGraw Hill, New York.

Vischer, J.C. (1989) *Environmental Quality in Offices*, Van Nostrand Reinhold, New York.

Williams, B. (1988) *Premises Audits*, Bulstrode Press, London.

Williams, B. (1994) *Facilities Economics*, Building Economics Bureau, Bromley.

Wilson, S. *et al* (1984) *Premises of Excellence*, Building Use Studies, London.

Wineman, J. (1986) *Behavioural Issues in Office Design*, Van Nostrand Reinhold, New York.

Worthington, J. and Konya, A. (1988) *Fitting Out the Workplace*, Architectural Press, London.

Wrennall, W. and Lee, Q. (1994) *Handbook of Commercial and Industrial Facilities Management*, McGraw-Hill, New York.

Wysocki, R. and Young, J. (eds) (1990) *Information Systems: Management Principles in Action*, John Wiley & Sons, New York.

Zeisel, J. (1981) *Inquiry by Design: Tools for Environmental Behaviour Research*, Cambridge University Press, Cambridge.

Zcithaml, V.A., Parasuraman, A. and Berry, L.L. (1990) *Delivering Quality Service*, Free Press, New York.

Zube, E.H. (1984) *Environmental Evaluation: Perception and Public Policy*, Cambridge University Press, Cambridge.

Index